VOLTAGE REFERENCES

Books of Related Interest from IEEE Press

Design of High-Performance Microprocessor Circuits
Anantha Chandrakasan, William Bowhill, and Frank Fox
2000   Hardcover   592pp   IEEE Order No. PC5836   0-7803-6001-X

Low-Power CMOS Design
Anantha Chandrakasan and Robert Brodersen
1998   Hardcover   644pp   IEEE Order No. PC5703   0-7803-3429-9

High-Performance System Design: Circuits and Logic
A volume in the IEEE Press Series on Microelectronic Systems
Vojin G. Oklobdzija
1999   Hardcover   560pp   IEEE Order No. PC5765   0-7803-4716-1

VOLTAGE REFERENCES
From Diodes to Precision
High-Order Bandgap Circuits

GABRIEL A. RINCÓN-MORA, PH.D.
Texas Instruments, Inc.
Dallas, Texas

Georgia Institute of Technology
Atlanta, Georgia

IEEE Press

JOHN WILEY & SONS, INC.

For ordering and customer service, call 1-800-CALL-WILEY

Library of Congress Cataloging in Publication Data is available.

ISBN 0-471-14336-7

10 9 8 7 6 5 4 3

In the blissful search for enlightenment!

CONTENTS

PREFACE

This textbook is a reference and tutorial on the design of integrated voltage references. The intent and focus of this book is to present a complete document that covers the conceptual history and the plethora of practical design issues behind integrated voltage references. The target audience is the circuit design community, from the novice digital and analog designer to the more experienced engineer.

Voltage references have always been an essential component of any system and consequently an important topic to explore. The last decades thrust toward higher and even total system integration has required all designers to be knowledgeable of this particular topic due to its mixed-signal implications (i.e., interface requirements and parametric considerations such as loading, output impedance, temperature coefficient, etc.) Bipolar, CMOS, and biCMOS designs and their pertinent issues are therefore discussed. As a result of the emergence of the portable battery-operated environment, low voltage and low power are key characteristics and consequently also included in the discussions.

The whole subject matter is divided into five chapters: *The Basics, Current References, Voltage References, Designing Precision Reference Circuits,* and *Considering the System and the Working Environment.* In the first chapter, the basic principles and components of reference circuits are defined and introduced. The following chapter deals with the design of current references, from basic CMOS PTAT circuits to complex biCMOS current generators. The third chapter takes the components introduced in the previous two chapters to illustrate how a voltage reference is designed, from basic zero-order to high-order circuits. Ultimately, practical design considerations conducive to the design of *precision* circuits such as process variants, loading effects, and operating environment are then explored in the fourth chapter. Chapter 5 brings closure to the topic by dealing with the issues of real-life applications, as they pertain to marketable integrated references, such as trim, package-shift effects, layout considerations, and characterization.

G.A.R.M.

Dallas, Texas

ACKNOWLEDGMENTS

First and foremost, I would like to thank God for the opportunity to make this work available to the community at large. By the same token, I am extremely thankful for the encouragement and support of my parents, Gladys M. and Gilberto Rincón, and my brother, Gilberto Alexei Rincón-Mora. They have always been a source of inspiration.

My deepest appreciation goes to Texas Instruments (TI) for its continued support throughout the years. I am also especially grateful for the encouragement and the advice of the following individuals: Professors Philip Allen and J. Alvin Connelly from Georgia Institute of Technology and Nicolas Salamina, David Briggs, and Ramanathan Ramani from Texas Instruments. I would also like to thank Buddhika Abesingha for his work and contribution to the study of package shift effects on bandgap references.

G.A.R.M.

Dallas, Texas

ACKNOWLEDGMENTS

LIST OF TABLES

LIST OF FIGURES

SUMMARY

Generally, references are an essential part of most, if not all, electrical systems. Although highly accurate precision references are usually desired, they are not always necessary. Rudimentary references are often sufficiently accurate to satisfy the demands of many applications. As a result, the complexity of the circuits varies from one design to the next. A simple reference may simply be a naturally existing voltage that does not change significantly with operating conditions. These references are zero order because there is no effort on the part of the designer to improve initial accuracy. Precision references, however, improve accuracy by attempting to cancel the linear (first-order) and nonlinear (second-order and higher-order) components of a given voltage. Due to more stringent design constraints, these higher-order circuits must also account for the effects of power supply voltage as well as electrical parasitic elements present. Some of these parasitic elements include resistor tolerance, resistor mismatch, transistor mismatch, leakage current, and package drift. These adverse effects are addressed at every level of design, be it circuit, layout, or trim.

Given an application whose accuracy requirement is low, though, a zero-order or first-order circuit is typically the least costly solution. Higher-order references are inherently more complex than lower-order circuits, thereby requiring more silicon area, more quiescent current flow, and possibly higher-input voltages. Though many applications still require lower-order references, curvature-corrected circuits are becoming increasingly popular in high-performance systems. The demand for high-performance references stems from several factors, ranging from reduced dynamic range (resulting from lower input voltages) to increased system complexity. Low power supplies, which result from battery operation and finer photolithography, and constant noise floors cause the effective dynamic range to be reduced.

Most precision references compensate the nonlinear temperature-dependent characteristics of a given naturally existing voltage by summing an artificially generated nonlinear component. The temperature dependence of this nonlinear term does not necessarily equal the nonlinearity of the naturally existing voltage defining the reference. For example, the correcting term of a bandgap circuit can be proportional to the square of the temperature while the nonlinearity of the diode voltage is actually proportional to a logarithmic expression of temperature ($T \ln T$, where T is temperature). Better performance is potentially viable when the correcting term is designed to match the nonlinearity of the temperature behavior of the freely existing voltage (i.e., match $T \ln T$ in a bandgap reference). Yet another genre of curvature correction attempts to exactly cancel the nonlinearity by deriving the correction component from the naturally existing voltage itself. This latter technique can potentially have even better performance than the matched version. Overall, the generation of these curvature-correcting components exploits the temperature characteristics of the devices available in a given process technology.

There are other aspects to designing a reference that are equally important as the temperature coefficient. The output stage, for instance, defines the output impedance of the circuit. In turn, this output characteristic partially dictates the circuits vulnerability to transient noise, sporadic or systematic in nature, as well as its ability to source and sink current. The design must mitigate the effects of steady-state changes and systematic noise present in the input supply voltage by designing for power supply rejection (PSR) and line regulation. Cascodes and pseudo-preregulated supplies perform this task well. The design must also account for process variations from die to die, wafer to wafer, and lot to lot. As a result, careful consideration is given to the design of the trim network. The system must be analyzed to determine the effects of circuit-block interface on the reference voltage. Systematic noise, for instance, may be injected through the substrate, power supply, and/or other circuits, thereby affecting the design of the circuit and the layout. Finally, the parametric limits of a design must be carefully characterized to ascertain and define its dependence on all the different factors, independent of one another. This step is important to fully comprehend the ramifications of the reference on the system.

Most precision references base their operation on the temperature dependence of a diode (base-emitter) voltage. However, the concepts and techniques presented in this book extend to any well-characterized naturally existing voltage. The essence of designing a temperature-independent reference involves the manipulation and the generation of several temperature-dependent components whose sum yields a low temporary coefficient (TC) output. This sum may be accomplished through currents and/or voltages. Furthermore, the techniques used to decrease the susceptibility of the bandgap reference to input power supply, output load, and process variants apply generally to the design of any reference. A very precise and commercially marketable reference is achieved when the circuit incorporates all these concepts into one design.

G.A.R.M

Dallas, Texas

CHAPTER 1

THE BASICS

Analog and digital circuits ultimately need a reference, be it voltage, current, or time. The reference establishes a stable point used by other subcircuits to generate predictable and repeatable results. This reference point should not fluctuate significantly under various operating conditions such as moving power supply voltages, temperature variations, and transient loading events. A few examples of circuit applications where references are intrinsically required are digital-to-analog converters, analog-to-digital converters, DC-DC converters, AC-DC converters, operational amplifiers, and linear regulators. These subsystems, of course, are the fundamental elements that make up cellular phones, pagers, laptops, and many other popular electronic products.

Though often neglected in discussions, references play a pivotal role in the design of integrated circuits. Their accuracy requirements, in fact, are more stringent in today's marketplace than ever before. The need for increased overall performance is the driving force behind this trend. In the future, as more complex and more compact systems emerge, this tendency toward high performance is expected only to increase. The prerequisite for high performance is partially exemplified by the growing demand for battery-operated circuits, which calls for high accuracy as well as low current overhead and low voltage operation. In addition to the increasing market demand for precision references, however, there is still, and will be, a necessity for crude low-precision voltage references. Among other functions, these low-precision references are indispensable for properly biasing circuit blocks.

The loading characteristics imposed on the reference by other circuits in the system determine the design constraints of the reference. For instance, a time-varying current load will require the reference to respond quickly to rapid transitions. In fact, depending on the speed of the transitions, a load capacitor may also be necessary to prevent the reference voltage from drooping excessively. As a result, not only will the circuit have to react quickly but it will also have to be able to drive relatively large capacitor values. These characteristics inevitably allude to a stable circuit with high closed-loop bandwidth response. For reference circuits, usually, the load-current transitions are not extensive (i.e., peak-to-peak currents of roughly less than 100 μA). If the load demands more than 100 μA, a voltage regulator is commonly used, which, by definition, regulates the output voltage against various loading conditions, including, most importantly, load-current changes. A voltage regulator is essentially a buffered reference. It is, in other words, a voltage reference cascaded with a unity-gain amplifier capable of driving higher currents.

Designing a voltage reference merits the scrutiny of several factors, which are mostly governed by the overall system. Temperature-drift performance is one of the most important issues with which to contend. For references where accuracy is paramount, a temperature-compensated reference is normally warranted. The general design approach for such a reference is to sum predictable, well-characterized, temperature-dependent components, voltages or currents, to yield a well-adjusted temperature-compensated response. For instance, a voltage that increases with temperature is summed with another that decreases with temperature to produce a temperature-independent voltage. The summing ratio, of course, is balanced such that their collective effect is low-voltage variations across the whole operating temperature range, which may span from -40 to 125 °C (commercial range). Ultimately, for precision references, summing parabolic temperature-dependent terms, like quadratic temperature-dependent components, are used to approximately cancel the undesired second-order effects exhibited by the diode voltage. The first step in the process, though, is to identify voltages and/or currents that are well characterized and that do not vary significantly with process. Typically, a *p-n* junction diode voltage is chosen for this purpose. The diode voltage is predictable, ± 2 to 5%, and is well characterized with temperature. Metal oxide semiconductors' (MOS) threshold voltages, although theoretically still viable, are less conducive for high accuracy since the initial accuracy is not as good as that of the p-n junction diode, typically ± 15 to 20%.

Integrated circuits, in general, are fabricated in a variety of process technologies ranging from standard bipolar and vanilla complementary metal oxide semiconductor (CMOS) processes to state-of-the-art silicon-on-insulator (SOI) and biCMOS technologies. As such, references may take one of several forms depending on the particular process for which they were designed. As the accuracy requirements of the applications increase, the complexity of the circuit also tends to increase to compensate for first-order, second-order, and even third-order parasitic effects. Similarly, references that must operate under low quiescent current flow and/or low supply voltage conditions are inherently more complex than those that need not meet such strict requirements. Whatever the level of complexity, though, all reference circuits stem from the same basic principles and components. Most process technologies, in fact, have a p-n junction diode available, which is the basis for most precision references. The techniques used to ultimately design the reference in different process technologies may differ slightly in practical terms but not in the conceptual sense. A bipolar reference, for instance, may use the base-emitter junction diode of an NPN transistor while a CMOS design will, more than likely, use the source-bulk junction diode of a p-type MOS transistor as the basic building block, which is none other than a p-n junction diode. Now, if another, more stable voltage or current were to become available, the techniques and the design approach would still remain essentially the same but under the guise of a different building block.

The performance of a reference is gauged by its variation and is described by its allowable operating conditions. The specifications of the reference include line regulation, temperature drift, quiescent current flow, input (power supply) voltage range, and loading conditions. Line regulation and temperature drift refer to the variations in reference voltage resulting from steady-state changes in power supply voltage and temperature. The typical metric used for variations across temperature is temperature coefficient (TC) and it is normally expressed in parts-per-million per degree Celsius (ppm/°C),

$$TC_{ref} = \frac{1}{Reference} \cdot \frac{\partial \, Reference}{\partial \, Temperature}, \qquad (1.1)$$

where the reference is either in volts, amps, or seconds. Overall accuracy is determined, primarily, by the inherent or initial accuracy of the reference and, secondarily, by line regulation and temperature-drift performance of the same, which is ultimately described by

$$
\text{Accuracy} = \frac{\Delta \text{Reference}_{IC} + \Delta \text{Reference}_{TC} + \Delta \text{Reference}_{LNR}}{\text{Reference}},
$$

(1.2)

where the subscripts IC, TC, and LNR refer to initial accuracy, temperature coefficient, and line regulation performance, respectively. Accuracy is specified in parts-per-million (ppm), percent, or bits [1]. For example, a 100-ppm, 2.5 V reference varies ± 0.25 mV ($\Delta V = 2.5V*100 \cdot 10^{-6}$), which is equivalent to a $\pm 1\%$ reference ($\% = 0.25$ mV $\div 2.5$ V). Similarly, the same reference is said to have 13 bits of accuracy while varying by the same amount ($\Delta V = 0.25$ mV ≤ 2.5 V $\div 2^{Bits}$ or Bits $\leq \log(2.5$ V $\div 0.25$ mV) $\div \log(2)$, where Bits is the maximum number of bits that still satisfies the relation). Load regulation is also often used to describe a reference and it refers to the effects of load on the reference, such as load current for the case of voltage references. Load regulation may be included in the metric for accuracy but it is most appropriately specified for regulator structures capable of handling a wider range of load currents, as discussed earlier.

This chapter, in particular, aside from introducing the topic, deals with the basics of designing a reference. The general topics discussed include diodes and current mirrors. They are the necessary building blocks used in the synthesis of any reference design, current or voltage reference. A design example is also presented to supplement the theory with a practical problem, within the context of a real working environment. The following chapter will then combine these circuit blocks to generate all sorts of current references. In the end, the current generators are then used to design voltage references.

1.1   THE DIODE

Figure 1.1 illustrates the current-voltage (I-V) relationship of a junction diode. The diode can be in the forward-biased, reverse-biased, or reverse-breakdown region. When the voltage across the terminals of

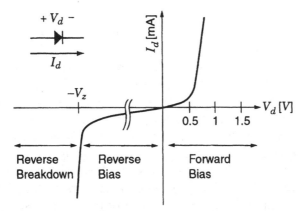

Figure 1.1 Typical *I-V* curve for a junction diode.

the diode (from anode to cathode) is between the breakdown voltage (denoted as $-V_z$) and approximately 0.5 V, the current flowing through the diode is significantly low. Once the current reaches the forward-biased region, it increases exponentially with the diode voltage,

$$I_D = I_S\left[\exp\left(\frac{V_D}{nV_T}\right) - 1\right],\qquad(1.3)$$

where I_D is the current flowing through the diode, I_S is the saturation current, V_D is the voltage across the diode, n is the ideality factor (a process-dependent constant), and V_T is the thermal voltage. The ideality factor is typically around 1. The thermal voltage is directly proportional to temperature,

$$V_T = \frac{kT}{q},\qquad(1.4)$$

where k is Boltzmann's constant (1.38×10^{-23} Joules/Kelvin), T is the absolute temperature in Kelvin degrees ($273.15 + °C$), and q is the magnitude of the electronic charge (1.602×10^{-19} Coulomb). At a temperature of 25 °C, the thermal voltage equals roughly 25.7 mV.

The reverse-biased current is approximately constant and nearly negligible. In other words, the device exhibits a large output resistance when reverse-biased. However, significant conduction into the cathode begins to occur when the voltage across the junction approaches the reverse-breakdown voltage $-V_z$. This conduction marks the onset of reverse breakdown. Operation in this region is not necessarily destructive but it may change some of the characteristics of the device. However, it is necessary to limit the reverse current flowing through the diode to prevent excessive power dissipation; otherwise, the device could experience irreversible damage. There are two different mechanisms responsible for diode breakdown: *avalanche multiplication* and *Zener tunneling*. Avalanche multiplication predominates at reverse-biased voltages of 7 V or greater. Zener tunneling is responsible for breakdown at reverse-biased voltages of 5 V or less. A combination of the avalanche and the Zener phenomena occurs for breakdown voltages between 5 and 7 V [2]. Irrespective of the mechanism, breakdown diodes are commonly referred as *Zener diodes*.

Zener tunneling is characterized by a negative temperature coefficient (TC). Avalanche multiplication, on the other hand, exhibits a positive TC [3]. As a result, breakdown diodes below 5 V have negative TCs, while those above 7 V have positive TCs. For example, the breakdown voltage of an emitter-base diode is typically 6 to 8 V with a TC of approximately $+2$ to 4 mV/°C. Generally, the temperature coefficient tends to become more positive (less negative) as the breakdown voltage increases, i.e., the 40 V base-collector Zener diode has a TC of roughly 35–40 mV/°C [3]. Zener diodes with breakdown voltages between 5 and 7 V, which have low TC values, are therefore appropriately used in voltage reference designs. This breakdown voltage range corresponds to the transition from Zener tunneling to avalanche multiplication.

1.1.1 Breakdown Region

The behavior of the diode in its reverse and breakdown regions is exploited in many reference applications. Since the diode is virtually an open circuit in its reverse region and a short circuit in its reverse-breakdown region, the device exhibits a voltage of $-V_z$ when current is flowing into the cathode terminal. The output impedance of the reference is simply the reciprocal of the slope of the I-V curve. As long as the diode remains in reverse breakdown, large changes in current produce small changes in voltage. In other words, the reference has a low output resistance. Diodes fabricated specifically for use in reverse breakdown are called *Zener diodes*. Their symbolic representation is illustrated in Figure 1.2, where V_z denotes the breakdown voltage (also called the *Zener voltage*) and current flows from the

Figure 1.2   Symbolic representation of a Zener diode.

cathode to the anode terminal of the diode. Typical output resistance values range from 10 to 300 Ω. This value includes the parasitic Ohmic resistances associated with both terminals of the actual p-n junction, diffusion resistance. It is the voltage span of most Zener diode references; 5 to 7 V Zener diodes have low TCs, which restricts them to relatively high voltage applications, power supply voltages greater than the respective breakdown voltage of the device. The temperature dependence of the familiar Zener voltage exhibits a positive temperature coefficient ranging from approximately $+1.5$ to 5 mV/°C.

1.1.2 Forward-Biased Region

The current-voltage (I-V) relationship in the forward-biased region is well represented by equation (1.3). For most of the practical operating range (currents ranging from a few to several hundred micro-amps), the diode voltage is approximately 0.6 V. This constancy results because of the exponential nature of the diode. Consequently, a forward-biased diode is useful in the generation of a repeatable and predictable voltage by forcing current into its anode. When operated in this mode, rearranging and differentiating equation (1.3) with respect to the diode current yields its output resistance (R_{out}), which is simply the reciprocal of the slope of the I-V curve,

$$R_{out} = \frac{\partial V_D}{\partial I_D} = \frac{nV_T}{I_S}\exp\left(-\frac{V_D}{nV_T}\right) \approx \frac{nV_T}{I_D}. \qquad (1.5)$$

It is important to note that the typical temperature dependence of this voltage is approximately -2.2 mV/°C!

The temperature dependence of a forward-biased diode is not linear, however. Its dependence on temperature is described by

$$V_D \approx V_{go} - \frac{T}{T_r}[V_{go} - V_D(T_r)] - (\eta - x)V_T \ln\left(\frac{T}{T_r}\right), \qquad (1.6)$$

where V_{go} is the diode voltage at 0 °K, T is the absolute temperature in degrees Kelvin (°K), $V_D(T_r)$ is the voltage across the diode at temperature T_r, η is a temperature-independent and process-dependent constant ranging from 3.6 to 4, and x refers to the temperature dependence of the current forced through the diode ($I_D = DT^x$, where D is a temperature-independent constant and x equals 1 for a proportional-to-absolute temperature current). The form of this relationship is a variant of the one presented by [4]. A current having a positive temperature coefficient (TC) is intuitively a good choice for a diode current. The positive TC can partially compensate for the negative TC of the diode voltage. The wisdom behind this choice is also obvious from the relation above since the logarithmic coefficient decreases as the order of the positive temperature-dependent current (x) increases. The diode relation is intrinsic for understanding and designing accurate references where the second-order components contained in the logarithmic term of equation (1.6) are compensated. As a result, a full characterization of the linear and quadratic dependence of the diode voltage is sometimes useful in the design process. Obtaining the Taylor series expansion of the logarithmic component and collecting the appropriate terms aptly describe its dependence to temperature, which is

$$
V_{BE} \approx \left[V_{go} + (\eta - x)V_{T_r} \right] - \left[\frac{V_{go} - V_{BE}(T_r) + (\eta - x)V_{T_r}}{T_r} \right] T
$$

$$
- \left\{ \frac{(\eta - x)V_{T_r}}{T_r} \left[T \ln\left(\frac{T}{T_r} \right) - T + T_r \right] \right\}. \tag{1.7}
$$

It is noted that the logarithm changes the coefficient of the linear component as well as that of the temperature-independent component. The derivation for equations (1.6) and (1.7) is presented in Appendix A.1 at the end of the chapter.

1.2 CURRENT MIRRORS

Current mirrors are widely used in basically all reference circuits. Consequently, a brief look at their operation and design considerations is justified. Figures 1.3 through 1.6 illustrate various bipolar and CMOS implementations of the current mirror, ranging from "simple" to regulated cascode mirrors. The basic function of a mirror is to take

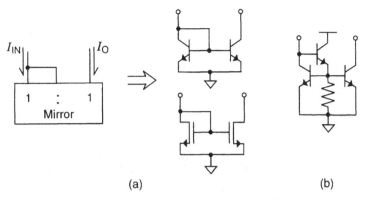

(a) (b)

Figure 1.3 Simple current mirrors.

an input current and duplicate a multiplied ratio of the same to another terminal. Intuitively, by looking at Figure 1.3(a), the input current charges the base node of the bipolar mirror until the sum of the collector current and the base current is equal to the input current, which occurs when the circuit reaches equilibrium. Well, given that the transistor sizes are the same and the output current is mostly a function of the base voltage, as described in equation (1.3) where the diode voltage is the base-emitter voltage, the output current is equal to the input current minus the base currents. Fortunately, the base currents are much smaller than the collector current, a beta factor smaller (β), which is between 70 to 100 times smaller. Consequently, the output current is roughly equal to the input current. Had the output device been twice the size of the input transistor, the output current would have been double the input current.

Figure 1.3(b) shows a version of the same current mirror where the approximation error introduced by the base current is reduced by another β factor since another NPN transistor is added. The resistor is not always necessary but it is added to ensure that current flows through the extra NPN transistor, thereby exhibiting a more stable and predictable transient response. The CMOS version of the mirror, also shown in Figure 1.3(a), does not require this so-called β helper since its input resistance is virtually infinite; its gate current is on the order of pico-amps. Although neglected in the intuitive analysis, the output current dependence to output voltage, collector or drain voltage, is a key parameter of the mirror. Inherently, transistors exhibit an Early voltage effect, also known as channel-length modulation for MOS devices. This variation in output current, more commonly described by its output resistance, is considered to be negligible in many integrated circuits but not so in the design of references, where small changes in current can be significant. Cascode mirrors, consequently, are often used since they produce more accurate results.

1.2.1 The Simple Mirror

The predominant source of error in the CMOS current mirror is channel-length modulation (λ),

$$I_D = I_{D\text{-sat}}(1 + \lambda V_{DS}) \tag{1.8}$$

where I_D is the drain-source current and $I_{D\text{-sat}}$ is the current once the device is in saturation. This basic MOS relation is used to derive the output current of the simple CMOS mirror (I_O), as depicted in Table 1.1. It is seen that the output current is not exactly equal to the input current; in fact, it is a function of both drain-source voltages. The bipolar circuit, as mentioned in the previous discussion, has an additional source of error: base current. Since

$$I_C = \beta I_B, \tag{1.9}$$

$$I_{\text{IN}} = I_{C1} + \frac{I_{C1} + I_{C2}}{\beta}, \tag{1.10}$$

and

$$I_C = I_S \exp\left(\frac{V_{BE}}{V_T}\right)\left(1 + \frac{V_{CE}}{V_A}\right), \tag{1.11}$$

where I_B is base current and V_A is Early voltage, the relationship of the output current to β and Early voltage effects is simply determined

TABLE 1.1. Output Current-Mirror Comparison Between Simple and Cascode Mirrors

	Simple Mirror—I_O	Cascode Mirror—I_O
CMOS	$I_{\text{IN}}\left(\dfrac{1 + \lambda V_{GS}}{1 + \lambda V_{DS}}\right)$	I_{IN}
bipolar	$I_{\text{IN}} \div \left[\dfrac{1}{\beta} + \left(\dfrac{1 + \lambda V_{BE}}{1 + \lambda V_{CE}}\right)\left(1 + \dfrac{1}{\beta}\right)\right]$	$I_{\text{IN}} \div \left[\dfrac{2}{\beta} + 1\right]$

by combining the above relations and collecting terms. The resulting equation is shown in Table 1.1. Connecting another NPN transistor, as shown in Figure 1.3(b), from the collector to the base of the input transistor minimizes the aforementioned β error at the cost, of course, of headroom limitations, two base-emitter voltages instead of just one. Overall, the output resistance of the simple mirror is comparable in both CMOS and bipolar technologies. The bandwidth is greater, the noise is lower, and the matching capability is better for the bipolar version, though. However, these benefits come at the expense of more current error resulting from β effects.

1.2.2 Cascode Mirrors

The evolution from the simple mirror to the cascode counterpart is readily apparent, shown in Figure 1.4. A couple of cascode devices are added to ensure that the drain-source or collector-emitter voltages of both input and output transistors are the same, thereby practically eliminating channel-length modulation and Early voltage effects. The voltage at the base or gate of these devices is simply a low-precision steady-state bias voltage, which may be generated by running some current through a resistor and/or diode-connected transistor. The resulting output resistance is therefore increased. In particular, it is increased by the product of the transconductance and the output resistance of the cascading device for the CMOS case and by a β factor for the bipolar circuit. Consequently, the much-improved input-to-output current relations shown in Table 1.1 result where the output current is essentially independent of drain-source and collector-emitter voltages.

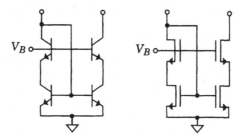

Figure 1.4   Cascode current mirrors.

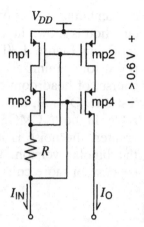

Figure 1.5 *P*-type cascode current-mirror example.

Design Example 1.1: Design a low-voltage, one-to-one, cascode current mirror (input current equals 10 μA, DC) whose output voltage is greater than 0.6 V below the positive supply voltage. The input and the output currents flow away from the mirror. Assume that a biCMOS process, with a minimum channel length for MOS devices of 1 μm, is to be used.

Since MOS devices are available and accuracy is best achieved with these transistors, *p*-type MOS transistors are chosen for the task. Figure 1.5 illustrates the circuit. Transistors mp1 and mp2 constitute the basic mirror while transistors mp3 and mp4 are the cascoding devices. Resistor *R* is simply used to generate a bias voltage for the gates of mp3 and mp4. Since the output voltage can swing up to within 0.6 V of positive supply V_{DD}, the sum of the saturation voltages across mp2 and mp4 must be less than 0.6 V. Since the mirror ratio is one-to-one, mp1 and mp2 must also be equal in size. For best results, mp3 and mp4 should also be the same size to ensure equal voltages at the drains of mp1 and mp2. The low overhead voltage of 0.6 V is arbitrarily chosen to be distributed equally between mp2 and mp4. Thus, choosing a saturation voltage (V_{sat}) of 0.25 V for all devices yields

$$V_{sat1} = V_{sat2} = V_{sat3} = V_{sat4} = \sqrt{\frac{2I_D}{K'(W/L)}}$$

$$= \sqrt{\frac{2I_{IN}}{K'(W/L)}} < 0.25\ V,$$

where K' is assumed to be 15 $\mu A/V^2$. The aspect ratio of all four devices is therefore chosen to be 25 $\mu m/\mu m$,

$$(W/L) > \frac{2I_{IN}}{K'(0.25^2)} = \frac{2(10\mu)}{(15\mu)(0.25^2)} = 21.3 \ \mu m/\mu m.$$

Critical device mp1 and device mp2 must match well for good overall performance of the mirror; hence, their channel length is chosen to be six times larger than the minimum allowed by the process to minimize channel-length modulation (i.e., $6\mu m$). The channel-length modulation effects of the cascoding devices on the output current are not as significant. As a result, their channel length is chosen to minimize area (i.e., 2 μm). Consequently, the dimensions (W/L) chosen for mp1 through mp4 are 150/6 $\mu m/\mu m$, 150/6 $\mu m/\mu m$, 50/2 $\mu m/\mu m$, and 50/2 $\mu m/\mu m$, respectively.

Finally, to ensure proper operation when the output voltage is within 0.6 V of the positive supply, the source-drain voltage of mp1 and mp2 is designed to be roughly 0.3 V, which would allow the output voltage to swing within 0.55 V of the positive supply (0.3 V + V_{sat}). The following loop equation is consequently used to ascertain the value of the resistor:

$$V_{SG1} + V_R - V_{SG3} - V_{SD1} = 0 = V_{SG1} + V_R - V_{SG3} - 0.3 \ V.$$

Since V_{SG1} is equal to V_{SG3} (same aspect ratio and equal current densities), they cancel and the voltage across the resistor is simply designed to be 0.3 V, which means R is 30 kΩ,

$$R = \frac{0.3 \ V}{I_{IN}} = \frac{0.3 \ V}{10\mu} = 30 \ k\Omega.$$

1.2.3 Regulated Cascode Mirrors

A current mirror is actually a single-stage amplifier with negative feedback. The input NPN transistor in Figure 1.3(a) is a common-emitter amplifier whose output, the collector terminal, is connected to its input, the base terminal, for negative feedback. The output current is therefore regulated against the input current. Similarly, a cascode device uses a common-gate or common-base amplifier stage to increase the output resistance of the overall circuit. However, its regulation performance may be further increased if higher open-loop gain

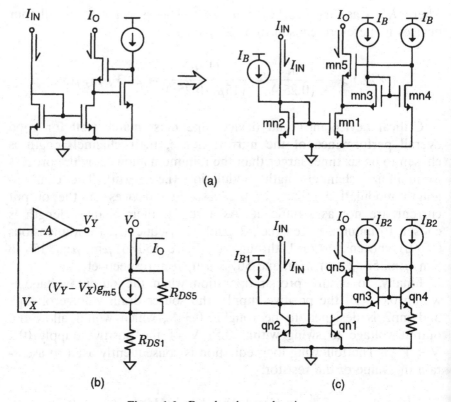

Figure 1.6   Regulated cascode mirrors.

were to be introduced in the cascode gain stage. Regulated cascode mirrors do just that. The output resistance of the current mirror is consequently boosted by a factor equal to that additional gain. Figure 1.6 shows some CMOS as well as bipolar implementations of regulated cascode mirrors [5, 6].

Figure 1.6(a), in particular, is a two-step evolution of the CMOS-regulated cascode circuit, from a high-voltage to a low-voltage circuit. In the final version, there is local feedback in addition to level shifting to assure a low voltage across the drain-source voltage of mn1 (V_{DS1}). Low voltage is desired to extend the working range of the mirror (i.e., the output voltage range for which the current is still regulated). The additional current source that is added to device mn2 (I_B) is used to match the current flowing through mn1, which is $I_B + I_O$, to avoid output current errors. Transistors mn3 and mn4 are designed to have a gate-source voltage difference equal to V_{DS1}, which is designed to be close to the saturation voltage of mn1 to maximize output voltage range.

Figure 1.6(b) shows the equivalent circuit model used to derive the output resistance of the mirror at hand. It is apparent from the circuit that the following four relations apply:

$$I_O = \frac{V_X}{R_{DS1}}, \tag{1.12}$$

$$V_Y = -AV_X, \tag{1.13}$$

$$V_O - V_X = [I_O - (V_Y - V_X)g_{m5}]R_{DS5}, \tag{1.14}$$

and

$$R_O = \frac{V_O}{I_O} = \frac{I_O R_{DSI} + (V_O - V_X)}{I_O}, \tag{1.15}$$

where g_m denotes transconductance and A is the additional gain associated with the cascode circuit ($A \approx g_{m4}R_{DS4}$). Recombining the aforementioned equations and collecting terms yields

$$R_O \approx R_{DS1}(Ag_{m5}R_{DS5}) = R_{DS1}(g_{m4}R_{DS4}g_{m5}R_{DS5}). \tag{1.16}$$

By equating A to 1, the relation for the regular cascode circuit is ascertained.

In the bipolar version of the same circuit, shown in Figure 1.6(c), a resistor is used to define the voltage across mirroring device qn1. More current error has now been introduced to the circuit, though. In particular, the culprits are qn3's and qn5's base currents (β errors). The following relation highlights the nature of these errors:

$$I_O\left(1 + \frac{1}{\beta}\right) + I_{B2}\left(1 - \frac{1}{\beta}\right) = (I_{IN} + I_{B1})\left(1 - \frac{2}{\beta}\right). \tag{1.17}$$

The error, of course, is minimized by appropriately sizing I_{B1} and I_{B2}. In the end, however, the complexity and the degraded frequency response of the regulated cascode mirror (loop-bandwidth response) limit its use. This trait encourages the designer to use, whenever possible, the simple and cascode mirrors in most analog design applications.

1.3   SUMMARY

Junction diodes are key elements in the design of references. They are not necessarily a requirement but they are certainly useful when seeking high performance at a reasonable cost. This trait arises because the p-n junction voltage has a relatively high initial accuracy. Additionally, the electrical characteristics are repeatable, predictable, and well characterized over a wide range of currents and temperatures. Consequently, most current and voltage references ultimately use these p-n junction diodes as the intrinsic building blocks in their respective designs.

Generally, references are an essential part of most, if not all, electrical systems. Although highly accurate precision references are usually desired, they are not always necessary. Rudimentary references are often sufficiently accurate to satisfy the demands of many applications. As a result, the complexity of the circuits varies from one design to the next. A simple reference may be a naturally existing voltage that does not change significantly with operating conditions, like the p-n junction diode. Precision references, however, improve accuracy by attempting to cancel the linear and nonlinear components of a given voltage. At this point, though, the basic tools with which references are designed have been explored (i.e., diodes and current mirrors). These are used in a variety of combinations to produce temperature-dependent currents and voltages, which eventually become key elements in the design of precision reference circuits. The next chapter deals with the generation of these temperature-dependent current references.

APPENDIX A.1   TEMPERATURE DEPENDENCE OF THE DIODE VOLTAGE

The collector current of a bipolar transistor exhibits an exponential relationship to the base-emitter voltage and is nominally expressed as

$$I_C = I_S \exp\left(\frac{V_{BE}}{V_T}\right),$$
(A.1.1)

where I_C is the collector current, I_S is the saturation current in the forward-active region, V_{BE} is the base-emitter (diode) voltage, and V_T is the thermal voltage. The effects of Early voltage are neglected for this derivation. Consequently, V_{BE} is derived to be

$$V_{BE} = V_T \ln\left(\frac{I_C}{I_S}\right),$$ (A.1.2)

The saturation current I_S is defined by the electron charge q (1.6 × 10^{-19} Coulomb), the intrinsic carrier concentration n_i (approximately 1.5 × 10^{10} cm^{-3} at 300 °K for silicon), the diffusion constant for electrons D_n, the emitter cross-sectional area A_e, the effective width of the base W_B, and the base-doping density N_A (assumed to be constant) [7],

$$I_S = \frac{qn_i^2 D_n A}{W_B N_A} = \frac{qn_i^2 \overline{D_n} A_e}{Q_B},$$ (A.1.3)

where Q_B is the number of doping atoms in the base per unit area of emitter and $\overline{D_n}$ is the average effective value of the electron diffusion constant in the base. The notation is consistent with NPN transistors but the theory also applies to PNP devices. The temperature dependence of the intrinsic carrier concentration and the average electron diffusion constant can be described by [8]

$$n_i^2 = AT^3 \exp\left(\frac{-V_{go}}{V_T}\right),$$ (A.1.4)

and

$$\overline{D_n} = V_T \overline{\mu_n},$$ (A.1.5)

where A is a temperature-independent constant, T is absolute temperature, V_{go} is the extrapolated diode voltage at 0 °K, and $\overline{\mu_n}$ is the average mobility for minority carriers in the base,

$$\overline{\mu_n} = BT^{-n},$$ (A.1.6)

where B and n are temperature-independent constants. The relationship for the saturation current I_S is more explicitly expressed by substituting equations (A.1.4) – (A.1.6) in (A.1.3), resulting in

$$I_S = \frac{q\left[AT^3\exp\left(\frac{-V_{go}}{V_T}\right)\right](V_T BT^{-n})A_e}{Q_B} \qquad \text{(A.1.7)}$$

or

$$I_S = CT^{(4-n)}\exp\left(\frac{-V_{go}}{V_T}\right), \qquad \text{(A.1.8)}$$

where C is a temperature-independent constant defined by all the constants in equation (A.1.7), such as q, A, B, A_e, Q_B, and k/q from the thermal voltage term ($V_T = kT/q$, where k is Boltzmann's constant: 8.62×10^{-5} eV/° K). Finally, the collector current can be assumed to have a temperature dependence whose behavior can be described by

$$I_C = DT^X, \qquad \text{(A.1.9)}$$

where D is a constant and x is an arbitrary number defined by the temperature dependence of the current forced through the collector; i.e., x is 1 for a proportional-to-absolute temperature (PTAT) current ($I_C \propto T^1$). Consequently, the temperature dependence of the base-emitter voltage can be reexpressed by substituting equations (A.1.8)–(A.1.9) in (A.1.2), resulting in

$$V_{BE} = V_T \ln\left[\frac{D}{C}T^{(x-(4-n))}\exp\left(\frac{V_{go}}{V_T}\right)\right]$$

$$= V_{go} + V_T \ln\left(\frac{D}{C}\right) - [(4-n) - x]V_T \ln T. \quad \text{(A.1.10)}$$

However, a more appropriate form of the base-emitter relationship, for the purpose of design, is its temperature dependence as a function of a reference temperature (T_r). This form can be derived by obtaining the relation of the base-emitter voltage (V_{BE}) at a reference temperature (T_r), solving for a constant, and substituting it back in equation

(A.1.10). The relation of V_{BE} at T_r is

$$V_{BE}(T_r) = V_{go} + V_{T_r} \ln\left(\frac{D}{C}\right) - [(4 - n) - x]V_{T_r} \ln T_r, \quad \text{(A.1.11)}$$

where V_{T_r} is the thermal voltage evaluated at the reference temperature T_r. At this point, the constant is derived to be

$$\ln\left(\frac{D}{C}\right) = \frac{V_{BE}(T_r) - V_{go} + [(4 - n) - x]V_{T_r} \ln T_r}{V_{T_r}}. \quad \text{(A.1.12)}$$

Now, equation (A.1.12) is substituted back in (A.1.10) to yield the well-known temperature dependence relationship of the base-emitter voltage,

$$V_{BE} = V_{go} - \frac{T}{T_r}[V_{go} - V_{BE}(T_r)] - [(4 - n) - x]V_T \ln\left(\frac{T}{T_r}\right).$$

$$\text{(A.1.13)}$$

It is often useful to develop the Taylor series expansion of the logarithmic term and substitute it back in equation (A.1.13). The purpose for the expansion is to more accurately design the cancellation of the linear as well as the curvature-correcting component of the bandgap reference. The process-dependent constant $(4 - n)$ is sometimes expressed as η with an approximate value between 3.6 and 4 [4]. The V_{BE} relationship can be rewritten as

$$V_{BE} = A + BT + Cf(T), \quad \text{(A.1.14)}$$

where A, B, and C are constants and $f(T)$ represents all the terms whose order is greater than 1 (i.e., $C_2T^2 + C_3T^3 + \cdots + C_nT^n$. Now the Taylor series expansion of the logarithmic term about the reference temperature T_r can be developed,

$$-(\eta - x)V_{T_r}\frac{T}{T_r}\ln\left(\frac{T}{T_r}\right)$$

$$= \frac{-(\eta - x)V_{T_r}}{T_r}\left[a_0 + \frac{a_1(T - T_r)^1}{1!}\right.$$

$$\left. + \frac{a_2(T - T_r)^2}{2!} + \cdots + \frac{a_n(T - T_r)^n}{n!}\right], \quad \text{(A.1.15)}$$

where the coefficients $a_0, a_1, \ldots, a_n$ are described by

$$a_k = \left. \frac{\partial^k \left[T \ln \left(\frac{T}{T_r} \right) \right]}{\partial^k T} \right|_{T=T_r}. \tag{A.1.16}$$

Equations (A.1.13), (A.1.15), and (A.1.16) are then used to derive the coefficients of equation (A.1.14) explicitly,

$$
V_{BE} \approx \left[V_{go} - \frac{(\eta - x)V_{T_r}}{T_r}(a_0 - a_1 T_r) \right]
$$

$$
- \left[\frac{V_{go} - V_{BE}(T_r)}{T_r} + \frac{a_1(\eta - x)V_{T_r}}{T_r} \right]T
$$

$$
- \left\{ \frac{(\eta - x)V_{T_r}}{T_r} \left[T \ln \left(\frac{T}{T_r} \right) - a_0 - a_1(T - T_r) \right] \right\} \tag{A.1.17}
$$

or

$$
V_{BE} = \left[V_{go} + (\eta - x)V_{T_r} \right] - \left[\frac{V_{go} - V_{BE}(T_r) + (\eta - x)V_{T_r}}{T_r} \right]T
$$

$$
- \left\{ \frac{(\eta - x)V_{T_r}}{T_r} \left[T \ln \left(\frac{T}{T_r} \right) - T + T_r \right] \right\}, \tag{A.1.18}
$$

which are variations of the relations offered by [8] and [9]. Given equations (A.1.14) through (A.1.18), higher-order terms like T^2, T^3, etc. can also be derived. It is noteworthy to mention that these higher-order terms will affect the lower terms since the expansion is done about temperature T_r, e.g., $(T - T_r)^2 = (T_r^2) - (2T_r)T + T^2$.

BIBLIOGRAPHY

[1] R. Kenyon, "A Quick Guide to Voltage References," *EDN*, no. 8, pp. 161–167, April 13, 2000.

[2] A.S. Sedra and K.C. Smith, *Microelectronic Circuits*. New York: Holt, Rinehart and Winston, 1987.

[3] A. Hastings, *The Art of Analog Layout*. New Jersey: Prentice-Hall, Inc., 2001.

[4] M. Gunawan et. al., "A Curvature-Corrected Low-Voltage Bandgap Reference," *IEEE Journal of Solid-State Circuits*, vol. 28, no. 6, pp. 667–670, June 1993.

[5] A.L. Coban and P.E. Allen, "A 1.75 V Rail-to-Rail CMOS Op Amp," *Proceedings IEEE International Symposium on Circuits and Systems*, vol. 5, pp. 497–500, 1994.

[6] M. Helfenstein et. al., "90 dB, 90 MHz, 30 mW OTA with the Gain-Enhancement Implemented by One- and Two-Stage Amplifiers," *Proceedings IEEE International Symposium on Circuits and Systems*, vol. 3, pp. 1732–1735, 1995.

[7] P.R. Gray and R.G. Meyer, *Analysis and Design of Analog Integrated Circuits*. New York: Wiley, 1993.

[8] Y.P. Tsividis, "Accurate Analysis of Temperature Effects in $I_c - V_{be}$ Characteristics with Application to Bandgap Reference Sources," *IEEE Journal of Solid-State Circuits*, vol. SC-15, no. 6, pp. 1076–1084, December 1980.

[9] G.M. Meijer et. al., "A New Curvature-Corrected Bandgap Reference," *IEEE Journal of Solid-State Circuits*, vol. SC-17, no. 6, pp. 1139–1143, December 1982.

CHAPTER 2

CURRENT REFERENCES

Most voltage references are derived from and are dependent on current references. The current reference does not need to be temperature-independent; however, its temperature drift must be well characterized and controlled. The absolute value of an integrated current reference is inherently variable because it is ultimately based on an integrated resistor, which typically exhibits a process variation of approximately 10 to 20%. A proportional-to-absolute temperature (PTAT) current, a current linearly proportional to temperature, is the most commonly used reference current. The popularity arises because the PTAT relation is practical, predictable, and linear over a wide range of currents. Additionally, amplifier with bipolar input pairs, which have a transconductance directly proportional to current and inversely proportional to temperature ($g_m = I_C \div V_T$), benefit from PTAT current references since the temperature dependence of the transconductance is effectively cancelled. Consequently, the gain-bandwidth product (GBW) of the amplifier becomes independent of temperature (GBW $\approx g_m/C$, where C is a frequency-compensation capacitor). Temperature-independent current references are also commonly used. These currents are usually derived from readily available constant voltage references or by properly combining temperature-dependent currents. Another useful current reference is the squared PTAT current (PTAT^2), where the current is proportional to the square of the temperature $(I \propto T^2)$, a quadratic dependence to temperature. This current is useful for constructing second-order voltage references, as will be discussed in the following chapter. This chapter discusses how these current references are designed. The underlying issues behind their practical implementation are also addressed within, of course, the constraints of mainstream processes like standard bipolar, vanilla CMOS, and biCMOS technologies.

2.1 PTAT CURRENT REFERENCES

2.1.1 Bipolar Implementations

Most PTAT current generators are constructed with bipolar transistors. Figure 2.1 illustrates how two simple topologies take advantage of the temperature dependence of the base-emitter voltage of an NPN transistor to produce a predictable temperature-dependent current. These circuits exploit the logarithmic relationship of the diode voltage (base-emitter voltage, V_{BE}), which is derived from equation (1.3) as

$$V_{BE} = V_T \ln\left(\frac{I_C}{J_S \text{Area}}\right), \qquad (2.1)$$

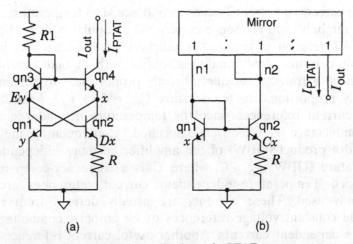

(a) (b)

Figure 2.1 Typical bipolar realizations of a PTAT current generator.

where I_C is the collector current of an NPN transistor and J_S is the saturation current density. As this equation shows, V_{BE} is directly proportional to the thermal voltage (V_T); in other words, it is directly proportional to absolute temperature (equation (1.4)). The temperature dependence of the logarithmic term is canceled by summing base-emitter voltages with equal collector currents, examples of which are shown in Figure 2.1. The following relation applies Kirchhoffs Voltage Law (KVL) to the loop containing transistors qn1–qn4, and resistor R of Figure 2.1(a):

$$I_{C2}R + V_{BE2} + V_{BE3} = V_{BE1} + V_{BE4}, \tag{2.2}$$

and substituting equation (2.1) into (2.2) yields

$$I_{C2} = I_{C4} = \frac{V_T}{R}\ln\left(\frac{I_{C4}I_{C1}DE}{I_{C2}I_{C3}}\right) = \frac{V_T}{R}\ln(DE)$$

$$= \frac{kT}{qR}\ln C = \frac{\Delta V_{BE}}{R} \equiv I_{\text{PTAT}}, \tag{2.3}$$

where $V_{BE1} - V_{BE4}$ are the base-emitter voltages of transistors qn1–qn4, respectively, T is absolute temperature, D and E are the emitter-area ratios of qn2/qn4 and qn3/qn1, V_T is the thermal voltage, ΔV_{BE} is the voltage difference of the base-emitter voltages, and base-current errors have been neglected. The constant C is the product of constants D and E and current I_{C3} is simply equal to current I_{C1}. As noted, I_{C2} is directly proportional to absolute temperature $(I_{C2} = I_{\text{PTAT}} \propto T)$. Resistor $R1$ is used as an imprecise current source $(I = (V_{DD} - 2V_{BE}) \div R)$. A significant attribute of this circuit is that the output current is independent of resistor $R1$'s current! The circuit, unfortunately, suffers from low power supply rejection ratio because the input current (I_{C3}) changes with power supply voltage since the voltage across $R1$ is dependent on the power supply voltage. As the input current varies, the effects of base-current errors also change, thereby adversely affecting I_{PTAT}.

Figure 2.1(b) shows an implementation where the currents through the NPN transistors are forced to equal each other by an active load mirror. The previous circuit is restricted to the use of an NPN cross-coupled quad to ensure proper operation and cancellation of currents (equation (2.3)). The simpler voltage loop is feasible in the latter version because the collector currents of the transistors around the loop are already equal. The output currents have the same relationship in both circuits, though. The drawback to the more accurate current generator, the latter circuit, is that it also has a stable zero-current state. Additional circuitry is required to prevent the circuit from approaching this state.

These PTAT generators can also be constructed with lateral PNP transistors; however, the topologies are not as robust as the NPN counterparts. This deficiency is mainly attributed to two factors, a low collector-current efficiency and a lightly doped base region. Collector-current efficiency for lateral PNP devices can be low especially if a highly doped deep buried layer is not available. The source for this low efficiency is the inherent existence of a parasitic vertical PNP transistor whose collector is connected to substrate (same base and emitter as the main lateral device). As a result, some of the emitter current flows to the substrate, thereby reducing the collector current efficiency of the lateral PNP device.

The second drawback arises because of the lightly doped nature of the base (n-well). Consequently, the current gain (forward-β) decreases rapidly with increasing collector current because of high-level injection. This phenomenon occurs when the minority-carrier density in the base-region becomes comparable to the majority-carrier density. High minority-carrier density causes the majority-carrier density to effectively increase so that charge neutrality is maintained, which, in turn, decreases the lifetime of minority carriers (there are more majority carriers with which to recombine) as well as increases the effective doping density of the base [1]. For proper operation, the minority-carrier density should be well below the majority-carrier level. High-level injection also occurs in NPN transistors, but at much higher current levels because the base has a higher doping density.

2.1.2 CMOS Implementations

PTAT current generators can also be designed with metal-oxide-semiconductor field-effect transistors (MOSFETs). In fact, using the same basic topologies of their bipolar counterpart, as shown in Figure 2.2 [2], they can be generated. The MOS devices, however, must operate in the subthreshold (weak inversion) region. This requirement arises

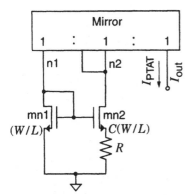

Figure 2.2 CMOS PTAT current generator.

because the drain current is exponentially dependent on the gate-source voltage only in subthreshold, which is the characteristic exploited by the circuit topology. Applying KVL to the voltage loop similarly derives the relationship of the output current,

$$I_{\text{out}} = \frac{V_T}{R} \ln \left[\frac{C(W/L)}{(W/L)} \right] = \frac{V_T}{R} \ln C, \qquad (2.4)$$

where (W/L) is the aspect ratio of the n-type device. One drawback to using subthreshold MOS devices is that leakage currents can overwhelm the drain current at moderately high temperatures [3]. Leakage currents increase as the temperature increases. As a result, the temperature range for which a subthreshold device is useful is sometimes limited.

Figure 2.3(a) illustrates a purely CMOS circuit architecture that does not rely on subthreshold operation to generate a PTAT current. In this case, p-n junction diodes are employed to generate the ΔV_{BE} voltage, or equivalently the ΔV_D voltage, relationship. Transistors mn1 and mn2 are used to force the voltages on nodes n1 and n2 to be equal. This equality occurs because both devices have the same gate-source voltage, a consequence of having equal dimensions and equal drain currents. Transistors mn1 and mn2 are the same size and are laid out to match well. The channel lengths of these devices must be large enough to prevent lambda effects (λ, channel-length modulation errors) from significantly degrading the accuracy of the PTAT current. If lambda effect is a problem, an operational amplifier can be used to ensure that the voltages on nodes n1 and n2 are equal, as

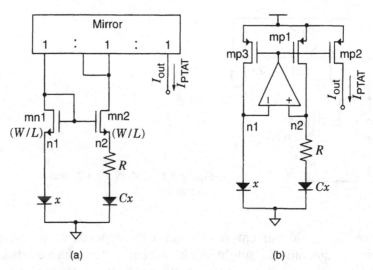

Figure 2.3   Other CMOS PTAT current generators.

shown in Figure 2.3(b). Transistor mp1 is used as the output device of the operational amplifier and transistors mp2 and mp3 are simply mirrors of mp1. The operational amplifier must exhibit a low input offset voltage, as this offset is a relatively high source of error. The resulting PTAT current relation for both circuits in the figure is the same as the one achieved by the circuits of Figures 2.1 and 2.2.

In standard p-substrate CMOS processes, there are three different p-n junctions available with which to create a diode, as illustrated in Figure 2.4. The p-epi/n-well and p-epi/$n+$ diffusion diodes, how-

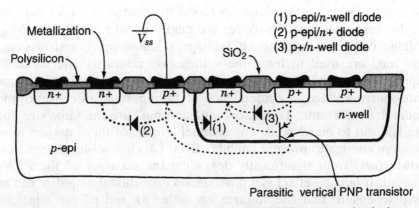

Figure 2.4   Diodes available in standard vanilla CMOS process technologies.

ever, have limited utility. This limitation is because the p-epi (substrate) is connected to the most negative potential, thereby eliminating the possibility of forward-biasing the junction. Furthermore, both diodes have significant series resistance because of the lightly doped nature of the well and the epitaxial layer. The third possible diode available in a standard CMOS process is created with the $p+$ diffusion normally used for PMOS devices and the n-well. The anode for this one ($p+$ diffusion), unlike the latter two, can be connected to a potential other than the most negative and would therefore be the one used for the PTAT current generator. Unfortunately, the physical layout of the device inherently contains a parasitic vertical substrate PNP transistor whose collector is the substrate. As a result, some of the diode current flows to the substrate, which constitutes an error for the circuit at hand. Furthermore, the device also suffers from parasitic Ohmic resistance, a consequence of using a lightly doped n-well.

2.2 STARTUP CIRCUITS AND FREQUENCY COMPENSATION

The current generators of Figures 2.1(b), 2.2, and 2.3 have an additional stable operating point, which is when I_{PTAT} is zero. A startup circuit (not shown) is required to prevent them from settling in this undesired state. A startup circuit ensures a nonzero current state by sourcing or sinking a minute amount of current to or from the current generator. This startup current is fed into a low-impedance node, such as nodes n1 and/or n2 in Figure 2.1. The startup current is normally significantly low and therefore does not need to be precisely controlled. Moreover, it can flow continuously or only when the circuit approaches its zero-current state. The latter is preferred for two reasons: (1) less power is consumed since there is less quiescent-current flow (especially important in battery-operated environments) and (2) the startup circuit does not perturb the current produced by the generator. This latter concern, for the case of continuous conduction, may be circumvented by implementing more complicated circuit topologies than those of Figures 2.1 (b), 2.2, and 2.3 where a small portion of current, startup current, flows into a nonsensitive current path.

2.2.1 Continuous-Conduction Startup Circuits

Figures 2.5 and 2.6 illustrate different startup circuit configurations that use junction field-effect transistors (JFETs) as imprecise current

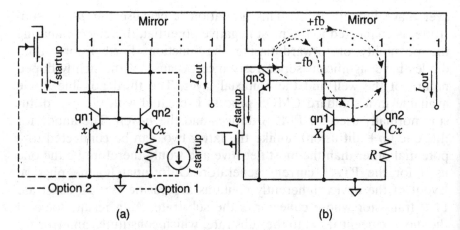

Figure 2.5 Continuous current startup circuit realizations.

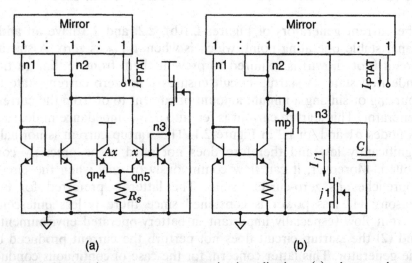

Figure 2.6 Noncontinuous current startup circuit realizations: (a) voltage-mode and (b) current-mode approach.

sources. Figure 2.5 depicts a configuration in which a continuous current is fed to a low-impedance node. In the case of the circuit of Figure 2.5(a), the startup current ($I_{startup}$) adversely affects I_{PTAT}. More specifically, I_{C1} differs from I_{C2} by an amount equal to $I_{startup}$ (error current). Thus, a small error is introduced into the logarithmic component of equation (2.3), i.e.,

$$I_{out} = \frac{V_T}{R} \ln\left[\frac{C(I_{C2} + I_{startup})}{I_{C2}} \right]. \tag{2.5}$$

The current $I_{startup}$ must therefore be much lower than I_{C1} and I_{C2} to prevent it from significantly degrading the PTAT nature of I_{out}. The circuit of Figure 2.5(b) does not suffer from this problem because it is designed in such a way that the startup current is not summed in a current-sensitive node. As depicted by the figure, I_{C1} and I_{C2} are equal if base current errors are neglected! Only the current through qn3 is affected by $I_{startup}$ and I_{PTAT} is independent of that current.

One advantage of this circuit is that I_{PTAT} will not change significantly with changes in power supply voltage. This immunity results because the collector voltages of qn1 and qn2 are approximately equal and are independent of the power supply. Parasitic Early voltage effects, which were disregarded in the design of the PTAT current, are minimized. The disadvantage of the circuit is its complexity, as it requires both more components and more current. In practice, low quiescent-current versions of this circuit tend to be more sensitive to power supply transients than the simpler circuit of Figure 2.5(a). In effect, power supply transients can momentarily interrupt the flow of current through this circuit more easily than the circuit of Figure 2.5(a). These interruptions affect the current produced by the generator and can cause severe system problems. An example of a system whose power supply is subject to large transients is an integrated switching power supply circuit. The systematic noise on the power supply as well as the nominal quiescent-current flow through the circuit will dictate the extent of the problem. Filter capacitors on the bases of qn1 and qn3 tend to alleviate these negative effects.

2.2.2 State-Dependent Startup Circuits

Another genre of startup circuits supplies, or sinks, current only when the circuit approaches the zero-current state. In other words, the state of operation is sensed. Figure 2.6(a) cites one such example, a voltage-mode approach [4]. Node n1 is used to ascertain the state of operation of the circuit. The circuit is operational if the voltage is greater than a diode voltage; otherwise, it is not. The voltage at node n3 is used as the reference point and is a forward-biased diode voltage ($V_{BE} > 0.5$ V). The differential pair composed of qn4 and qn5 compares the reference point to the signal at node n1. As a result, if the voltage at node n1 is below the reference point, the circuit is approaching the zero-current state, thereby forcing qn5 to pull current from node n2. On the other hand, if the current generator circuit is in its proper operational mode, the voltage at node n1 is roughly equal to that of reference node n3. Since transistor qn4 is "A" times larger than qn5, most of the current will flow through qn4 and not through qn5. The value of resistor Rs and the area-ratio between qn4 and qn5 determines the magnitude of the error current associated with the startup circuit during normal operating conditions. Circuits of this sort work well but generally tend to be complex and therefore present more risk with regard to overall yield performance (number of units in a given lot that meet all specifications).

State-dependent startup blocks can also take the form of current-mode circuits. These circuits tend to be robust as well as conducive for low-voltage operation. Figure 2.6(b) illustrates a current-mode circuit implementation. Unlike the voltage-mode counterpart, the state of the circuit is ascertained by monitoring a current, not a voltage. The current through the PTAT core circuit, in particular, is monitored and mirrored to node n3, a high-impedance node. This current is essentially compared to the current sunk by junction field-effect transistor (JFET) $j1$. Device $j1$ is configured as a constant-current sink whose value is designed to be somewhere between I_{PTAT} and zero, throughout the whole temperature range as well as over process variations. As a result, when the circuit is in the off state, n3 is pulled low; this causes transistor mp to sink current from node n2, which starts the circuit. Once the circuit starts (i.e., I_{PTAT} is greater than I_{j1}) node n3 is pulled back up high, which eradicates the gate drive and therefore prevents any current from flowing through device mp.

The disadvantage of current-mode startup circuits, however, is their vulnerability to transient noise. This noise, unfortunately, is prevalent in many ICs where complex systems have been integrated onto a single die. Transient noise, injected through the substrate, for instance, may alter the state of node 3, the high-impedance node. Capacitor C is added for this reason, to slow down voltage transitions on n3 and thus filter transient noise. The ultimate result of an insufficiently robust circuit, though, is a transient fluctuation of I_{PTAT}, which may be perceived, at the system level, as extraneous instabilities emanating from the reference and whatever other circuits are dependent on I_{PTAT} (as a biasing element). The extent of the vulnerability is a function of the value of I_{j1} and how low it is as well as the value of capacitor C. Consequently, increasing the values of capacitor C and current I_{j1} (or whatever constant bias current is used in the startup circuit) mitigates the undesired, aforementioned transient effects. In many mixed-signal environments, a combination of both voltage-mode and current-mode approaches is warranted for reliability in a noisy environment and predictability.

2.2.3 Frequency Compensation

Many of these circuits also require frequency compensation. The necessity arises because of the positive and the negative feedback loops in the circuit. In fact, two conflicting intertwined loops are often encountered. The circuit of Figure 2.5(b) illustrates one such example. The negative feedback path goes from the base of qn3 to the collector of qn3, which, in turn, goes to the base of qn1/qn2 through the mirror device defining I_{C1} (introduces one polarity inversion) and back to the base of qn3 through qn2. The positive feedback path, on the other hand, goes from the base of qn3 to the collector of qn3 and back to the base of qn3 through the mirror transistor defining I_{C2}. The criterion for this loop to be stable is that the positive feedback gain must be lower than the negative feedback gain at every frequency in the spectrum. The basic design approach, then, is to lower the bandwidth of the positive feedback loop to the point where the above requirement is ensured. This design process is typically iterative in nature and may take several simulation runs to determine the best placement and size of the compensating filters.

Stability of the circuit is analyzed by breaking each loop, thereby ascertaining the open-loop frequency response. In addition to the requirement of having the positive feedback gain be lower than the negative feedback gain at all frequencies, the circuit needs to have enough phase margin to ensure stability over process corners and temperature extremes. Ultimately, the robustness of the circuit is tested with a transient response analysis. In this technique, a node voltage or a current is transiently perturbed in simulations (e.g., a 1–100 ns current or voltage impulse on a particular node). If the circuit recovers and its output settles, then the circuit is less liable to become unstable. This test should be performed after all AC analyses have satisfied the requirements of stability. Unfortunately, satisfying the AC analysis requirements does not guarantee stability. This uncertainty arises because some of the characteristics usually change when the loop is broken, such as the loading conditions for the broken node. For instance, if the loop in the circuit of Figure 2.5(b) is opened at the base of qn3, then the collector of qn2 will not see the load-capacitance and the load-resistance of the base of qn3. The designer may replicate transistor qn3 or employ a similar impedance-matching technique to circumvent this phenomenon. In the end, however, only the transient analysis will be able to gauge the closed-loop stability characteristics of the circuit!

2.3 CTAT CURRENT REFERENCES

Another temperature-dependent current that is useful to generate in reference circuits is the complementary-to-absolute temperature (CTAT) current [5], the complement of PTAT. A CTAT dependence is often used in conjunction with curvature correction schemes for second-order voltage references as well as for first-order current references. *Second order* refers to the cancellation of linear and quadratic terms, with respect to temperature, while *first order* refers to the cancellation of only the linear term. This cancellation will become apparent in the following sections and in the following chapter. The CTAT component is generated practically by forcing a base-emitter (diode) voltage across a resistor and by mirroring the current flowing through the resistor elsewhere in the circuit. In essence, the strong linear component of the base-emitter (diode) voltage (roughly -2.2 mV/°C) is used. Figure 2.7 exemplifies two simple embodiments of this concept; a bipolar and a CMOS version. The output current (I_{CTAT}) is described as

$$I_{out} = \frac{V_{BE}}{R} \equiv I_{CTAT},$$ (2.6)

where V_{BE} is the base-emitter (diode) voltage across transistor qn. Sink current I_x can be designed to be any value as long as it is greater in magnitude than I_{CTAT} throughout the whole operating temperature range. This requirement is to ensure that transistor qn operates in the forward-active region. Current I_x should be PTAT in nature, if possible, to mitigate the nonlinear effects of the logarithmic component of V_{BE} (equation (1.6)). The circuit of Figure 2.7(b) uses an

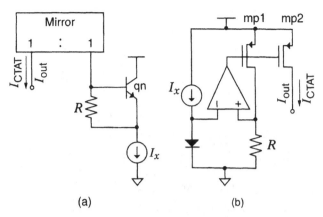

Figure 2.7   CTAT current generators.

operational amplifier to force a diode voltage across resistor R. This version is more compatible with CMOS technologies where bipolar transistors are not available. In this case, a $p+$ diffusion$/n$-well diode (Figure 2.4) can be used. Transistors mp1 and mp2 can be replaced with PNP devices if the process and/or application demands it. Transistors mp1 and mp2 are used only as mirroring devices.

2.4   TEMPERATURE-INDEPENDENT CURRENT REFERENCES

Summing PTAT and CTAT currents generates temperature-independent currents, first order. As the name implies, their sum can yield a quasi-temperature-independent current, if they are appropriately proportioned. This sum would create a first-order temperature-independent current. The simplest way of generating this current is by summing a PTAT current with the CTAT current of Figure 2.7(a). This addendum to the circuit is simple and effective. Another version, which incorporates the circuit that generates the PTAT current into the same design, is shown in Figure 2.8. In this configuration, the PTAT current is generated by the loop composed of qn1, qn2, and $R1$ (ΔV_{BE} loop) and the CTAT current is generated by qn2 and $R2$. Resistor $R3$ is added to ensure that the collector currents flowing through qn1 and qn2 are PTAT and equal; thus, $R3$ is designed to match $R2$. These circuits assume that the temperature coefficient (TC) of the resistors is low. Circuit techniques can be formulated to use the TC of the resistors to help shape the temperature dependence

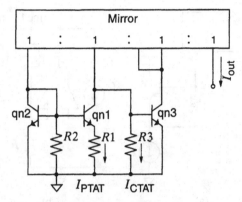

Figure 2.8 Quasi-temperature-independent current source.

of a particular current. Yet another simple method of generating a temperature-stable current is by forcing a temperature-independent voltage reference across a resistor. The premise, in this case, is again that a low-TC resistor is available but, more importantly, that a reference voltage already exists! As a result, this latter technique is restricted only to certain applications.

2.5 PTAT² CURRENT GENERATORS

2.5.1 Bipolar Implementations

A current component that is proportional to the square of the temperature (PTAT²), quadratic term, is also useful in implementing precision references, in particular, second-order curvature-corrected voltage references. Figure 2.9 shows one bipolar realization of the PTAT² current generator. In this circuit, a PTAT current and a first-order temperature-stable current is used in a base-emitter loop to produce the desired PTAT² current. Summing a CTAT and a PTAT current, as previously discussed, generates the temperature-independent current. The intrinsic voltage loop is composed of transistors qn1, qn2, qn3, and qn4. The resulting output current is derived by summing the voltages in the loop (applying KVL) and substituting equation (2.1) for each base-emitter voltage, thereby yielding

$$V_T \ln \left[\frac{I_{out}\left(\dfrac{I_{PTAT}}{A} + I_{CTAT}\right)}{I_{PTAT} I_{PTAT}} \right] = 0 \qquad (2.7)$$

or

$$I_{\text{out}} = \frac{I_{\text{PTAT}}^2}{\left(\dfrac{I_{\text{PTAT}}}{A} + I_{\text{CTAT}}\right)} \approx \frac{I_{\text{PTAT}}^2}{K} \equiv I_{\text{PTAT}^2}, \qquad (2.8)$$

where A and K are constants. Another straightforward technique of generating I_{PTAT}^2, though more complex in terms of transistor count and area overhead, is by using a Gilbert Multiplier and a PTAT voltage (ΔV_{BE}) as the two differential inputs of the same. Still another method is to simply force a PTAT current to flow through a resistor that exhibits a PTAT dependence (i.e., a metal resistor). This method, however, is dependent on the technology and is therefore implemented only if the process and application allow.

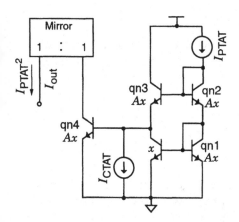

Figure 2.9 Bipolar PTAT² current generator.

2.5.2 CMOS Implementations

Similarly, the Gilbert Multiplier, as well as the positive TC resistor technique, can be used to generate PTAT² currents in a strictly CMOS environment. Figure 2.10, however, illustrates a vanilla CMOS realization of the PTAT² current generator [2]. This circuit takes advantage of the characteristics of MOS devices in subthreshold (weak inversion) and above-threshold (strong inversion) regions to create a current that is proportional to the square of the absolute temperature. In particular, transistors mn1 and mn2 operate in subthreshold while mn3 and mn4 operate in strong inversion. The key voltage loop in this circuit is composed of mn1, mn2, and mn4. The voltage across the

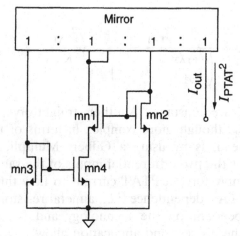

Figure 2.10 CMOS PTAT^2 current generator.

drain and source of mn4 is a ΔV_{GS} dictated by the characteristics of operation in weak inversion (behavior similar to a ΔV_{BE} voltage),

$$I_{out} = \frac{V_{GS2} - V_{GS1}}{R_{DS4}} = \frac{V_T}{R_{DS4}} \ln\left(\frac{\beta_1}{\beta_2}\right). \tag{2.9}$$

where V_{GS1} and V_{GS2} correspond to the gate-source voltages of mn1 and mn2, β_1 and β_2 refer to the products of the transconductance parameter (K') and the aspect ratios (W/L) of mn1 and mn2, and R_{DS4} is the output resistance of mn4. Since transistor mn4 is operating in strong inversion and, at the same time, is in the triode region, the following relation applies:

$$R_{DS4} \approx \frac{1}{\beta_4(V_{GS4} - V_{tn})} = \frac{1}{\beta_4(V_{GS3} - V_{tn})} = \frac{1}{\beta_4\sqrt{\frac{2I_{out}}{\beta_3}}}, \tag{2.10}$$

where V_{GS3} and β_3 correspond to mn3 and V_{tn} is the threshold voltage of n-type MOS devices. If equation (2.10) is substituted back into equation (2.9), solving for I_{out} reveals the temperature-squared dependence of the current, i.e.,

$$I_{out} = \beta_4 \sqrt{\frac{2I_{out}}{\beta_3}} \, V_T \, \ln\left(\frac{\beta_1}{\beta_2}\right)$$

or

$$I_{out} = \frac{2\beta_4^2}{\beta_3} V_T^2 \left[\ln\left(\frac{\beta_1}{\beta_2}\right)\right]^2 \equiv I_{PTAT^2}. \tag{2.11}$$

Design Example 2.1: Design an integrated version of a PTAT and a PTAT² current generator assuming that the power supply voltage is 5 V, the process-dependent constant η is 4, and bandgap voltage V_{go} is 1.2 V. Assume a biCMOS process is to be used.

The first section to be designed is the PTAT generator block. This block is composed of mp1–mp3, qn1–qn3, and R_{PTAT} in Figure 2.11. Transistor qn2 is chosen to have an area equal to four times larger than the minimum possible size for an NPN device. A nonminimum device size is chosen to maximize the matching capabilities between transistors qn2 and qn3. The current density should also be taken into account but will not usually be a limiting factor. Now, the current through qn3 is chosen to be 10 μA. Lower currents can be chosen; however, low current cells are more susceptible to substrate and power supply noise injection. Noise susceptibility is especially important in integrated mixed-signal designs. The size of qn3 needs to be larger than qn2 and, for this particular design, it is chosen to be 8 times larger than qn2. The ratio should not be any larger than 10 to

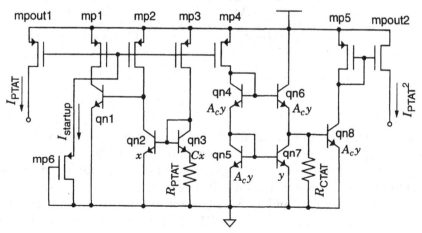

Figure 2.11 Example of a PTAT and a PTAT² current generator.

12. This restriction is because the matching capabilities start to degrade when the area-spread becomes large. As a result, resistor R_{PTAT} is simply

$$R_{PTAT} = \frac{V_T \ln C}{I_{PTAT}} = \frac{(25.8 \text{ m}) \ln 8}{10 \ \mu A} \approx 5.360 \ \Omega.$$

Transistor qn1 is only a feedback device and is therefore chosen to be minimum size. Transistors mp1–mp4 and mpout1 compose a mirror, which is chosen to source equal currents throughout. The PMOS transistors aspect ratios are designed to maximize accuracy (i.e., $W/L = 100/25$). The long channel length is chosen to minimize lambda effects (channel-length modulation errors) and the long width is chosen to maximize matching performance. The saturation voltage $(V_{GS} - V_t$ or $V_{GSt})$ is designed to ensure that the devices do not operate in the subthreshold region (i.e., $V_{GSt} > 150$ mV). Matching for devices operating in subthreshold is not as good as in strong inversion. A variation in the threshold voltage is a bigger percentage of the total gate-source voltage for a subthreshold device than for the same transistor in strong inversion. Transistor mp6 is used as a continuous-current startup device sinking some current from transistor mp1. Its current must not exceed or even approach the magnitude of the PTAT current; otherwise, transistor qn1 would be starved of current. Thus, a current of approximately 0.5 μA is chosen,

$$V_{SGt6} = V_{SG6} - |V_{tp}| = (5 \text{ V} - V_{SG1}) - |V_{tp}| = 5 \text{ V} - V_{SGt1} - 2|V_{tp}|$$

$$\approx 3.6 \text{ V} - \sqrt{\frac{2(10 \ \mu A)}{15\mu(100/25)}} \approx 3 \text{ V} \equiv \sqrt{\frac{2(0.5 \ \mu A)}{15\mu(W/L)_{mp6}}}$$

or

$$(W/L)_{mp6} = \frac{1}{135},$$

where the transconductance parameter K' is assumed to be 15 $\mu A/V^2$ and the threshold voltage $|V_{tp}|$ is assumed to be 0.7 V.

The current through transistor qn6 must be temperature independent (first-order approximation). This current is therefore designed to be the sum of a PTAT and a CTAT current. By imposing a base-emitter voltage across resistor R_{CTAT}, a CTAT current is generated. If the logarithmic term of the base-emitter voltage is neglected, then, from equation (1.7),

$$I_{\text{indep}} = I_{\text{PTAT}} + I_{\text{CTAT}}$$

$$\approx \frac{I_{\text{PTAT}}(T_r)}{A_c} \cdot \frac{T}{T_r}$$

$$+ \frac{\left[v_{go} + (\eta - 1)V_{T_r} \right] - \left[V_{go} - V_{BE}(T_r) + (\eta - 1)V_{T_r} \right]\dfrac{T}{T_r}}{R_{\text{CTAT}}},$$

where I_{indep} is the temperature-independent current, A_c a constant defined by the mirror-ratio of qn5 and qn7, and V_{T_r} is the thermal voltage at room temperature (300 °K). For I_{indep} to have a zero temperature coefficient (TC), the TCs of the PTAT and the CTAT currents must be equal and opposite. In other words, the first derivative of I_{indep} must be zero,

$$\frac{\partial I_{\text{indep}}}{\partial T} = \frac{I_{\text{PTAT}}(T_r)}{T_r A_c} - \frac{\left[V_{go} - V_{BE}(T_r) + (\eta - 1)V_{T_r} \right]}{T_r R_{\text{CTAT}}} \equiv 0$$

thus,

$$R_{\text{CTAT}} \equiv \frac{\left[V_{go} - V_{BE}(T_r) + (\eta - 1)V_{T_r} \right]}{I_{\text{PTAT}}(T_r)} A_c$$

$$\approx \frac{(1.2 - 0.6 + 77.4\text{m}) \text{ V}}{10 \ \mu\text{A}} 2 \approx 136 \ k\Omega,$$

where V_{BE} is assumed to be 0.6 V at room temperature while constant A_c is chosen to be two (for simplicity as well as for matching). At this point, the circuit simulator is used to fine-tune R_{CTAT} to the point where I_{indep} is properly centered. This tuning is necessary because Early voltage, channel-length modulation effects, nonlinearity of the diode voltage, and TC of the resistors were disregarded in the analysis. As a result, the linear TCs of the PTAT and the CTAT currents will be close but not exactly equal. In the end, the output PTAT^2 current is

$$I_{PTAT^2} = \frac{I_{PTAT}I_{PTAT}}{\left(\dfrac{I_{PTAT}}{A_c} + I_{CTAT}\right)} \approx \frac{10\mu10\mu}{\left(\dfrac{10\mu}{2} + \dfrac{0.6}{136k}\right)} \cdot \frac{T^2}{(300\ ^\circ\text{K})^2}$$

$$\approx 10.6\frac{T^2}{(300\ ^\circ\text{K})^2}\,\mu\text{A},$$

assuming that the mirror composed of mp5 and mpout2 has a ratio of 1 : 1. The aspect ratios of these devices are chosen in the same fashion as those of mp1 through mp4 (i.e., 100/25).

2.6 SUMMARY

Not only are current references used throughout integrated circuits in general but they are also intrinsic components of voltage references, which ultimately dictate the overall accuracy of the respective systems. Proportional-to-absolute temperature (PTAT) current references, for instance, are useful in generating reliable and predictable current sources. They are even used to temperature-compensate the gain-bandwidth product of bipolar-input amplifiers. Temperature-independent currents, of course, are also useful to predictably bias circuits across the temperature range with fixed current densities. Moreover, these currents, as well as complementary-to-absolute temperature (CTAT) currents, are used together to generate more complex temperature-dependent currents, such as currents with quadratic dependence to temperature, PTAT^2. A PTAT^2 current is especially useful when canceling the quadratic dependence of the diode voltage. In the end, though, a combination of these currents is used in precision references to cancel linear, quadratic, and higher-order temperature-dependent effects. The next chapter discusses exactly how these high-order references are developed and designed by using the current reference circuits described in this chapter.

BIBLIOGRAPHY

[1] P.R. Gray and R. G. Meyer, *Analysis and Design of Analog Integrated Circuits*. New York: Wiley, 1993.

[2] H.J. Oguey and D. Aebischer, "CMOS Current Reference Without Resistance," *IEEE Journal of Solid-State Circuits*, vol. 32, no. 7, pp. 1132–1135, July 1997.

[3] P.E. Allen and D.R. Holdberg, *CMOS Analog Circuit Design*. New York: Holt, Rinehart and Winston, 1987.

[4] G.A. Rincon-Mora and P.E. Allen, "A 1.1 V Current-Mode and Piece-wise-Linear Curvature Corrected Bandgap Reference," *IEEE Journal of Solid-State Circuits*, vol. 33, no. 10, pp. 1551–1554, October 1998.

[5] J.H. Huijsing, et al., *Analog Circuit Design*. The Netherlands: Kluwer Academic Publishers, 1996.

CHAPTER 3

VOLTAGE REFERENCES

Diodes, current mirrors, and, ultimately, current references comprise the necessary building blocks used in the synthesis of most, if not all, voltage reference topologies. Voltage references usually have more visibility and exposure than current references because their accuracy is typically higher and their outputs are more predictable. Precise, integrated current references are difficult circuits to design. This inaccuracy stems from the tolerance of integrated resistors, on the order of 10 to 20%, which, of course, directly translates to the overall tolerance of the current reference since the current is usually derived from a voltage across a resistor. proportional-to-absolute temperature (PTAT) and complementary-to-absolute temperature (CTAT) currents, for instance, are inversely proportional to resistors and they themselves are intrinsic elements in the generation of $PTAT^2$ currents (Chapter 2). The purely CMOS version of the $PTAT^2$ shown in section 2.5.2, on the other hand, is not dependent on a resistor but it is similarly dependent on the output resistance of a transistor, which also varies substantially with process. Trimming the current circumvents this inaccuracy; however, a significant number of bits is required to achieve a high level of accuracy. A voltage reference, since it is inherently more accurate, requires less trimming bits to achieve similar performance. The diode voltage, which is the basic building block of most voltage references, typically has a tolerance of roughly $\pm 2\%$ under a given collector current [1]. As a result, most of the theory and applications discussed in this chapter will be based on the diode or, equivalently, the base-emitter voltage of an NPN transistor. However, the techniques themselves also apply to any well-characterized voltage that a given process technology may offer.

A voltage reference can be categorized into different performance levels (i.e., zero order, first order, or second order). The zero order, as the name implies, is the most rudimentary. This reference can have a temperature-drift performance extending from 1.5 to 5 mV/°C. These types of references are typically not temperature-compensated; in other words, there is no effort on the part of the designer to improve the existing tolerance of the given voltage, which could be derived from a Zener or a forward-biased diode. On the other hand, first-order references are temperature-compensated. The first-order term of the polynomial relationship with respect to temperature is effectively canceled. A Taylor-series expansion of the voltage to be temperature compensated, which can be any voltage, is useful in designing for the cancellation of the first-order term and higher-order components. An example of this Taylor expansion is partially demonstrated in equation (1.7) of Chapter 1 for the diode voltage and fully derived in Appendix A.1 from Chapter 1. In any case, first-order voltage references exhibit a temperature drift ranging from 50 to 100 ppm/°C. There are some applications like high-performance data converters and low-voltage power supply systems that require even more accuracy than what a first-order voltage reference can supply. Second-order as well as higher-order references are used for this end. They typically vary less than 50 ppm/°C. These circuits, as their name implies, compensate for linear and one or more higher-order temperature components.

This chapter discusses the theory and issues, with respect to temperature, that surround the design of a voltage reference, be it first-order, second-order, or even a higher-order circuit. The use of the different building blocks developed in previous chapters is illustrated and discussed within the context of a practical voltage reference design. Examples of state-of-the-art designs are also cited, examined, and gauged against one another for their relative performance. Practical examples are also included in the text to supplement the discussion with "real-life" circuits.

3.1 ZERO-ORDER REFERENCES

3.1.1 Forward-Biased Diode References

The simplest and most financially economical method of generating a voltage reference is by forcing current to flow through a *p-n* junction

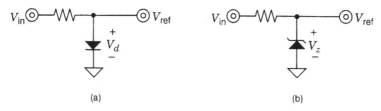

(a)   (b)

Figure 3.1   Diode voltage references.

diode. The circuit in Figure 3.1(a) exemplifies one simple realization. The resistor can be replaced with a current source if the system where the design lies has one available. It can also be replaced by a junction field-effect transistor (JFET) to optimize area and current overhead. This reference will have a temperature-drift performance of approximately -2.2 mV/°C. The accuracy performance is degraded if the input voltage changes and/or if its output is used to drive load currents. More specifically, the biasing current of the diode changes with input voltage variations, thereby causing, as a result, deviations in the output voltage. Similarly, the load-current degradation results because the output voltage exhibits a load-regulation drift when subjected to load-current changes, as described in equation (1.5) from Chapter 1.

3.1.2 Zener References

Another common realization of a voltage reference circuit is through the use of a Zener diode and a resistor [2, 3] as shown in Figure 3.1(b). If current is forced to flow into the cathode of a diode, the diode goes into the reverse-breakdown region. In this mode of operation, significant load-current changes cause nearly negligible fluctuations in diode voltage. As a result, a low output resistance at the cathode arises, varying typically from 10 to 300 Ω. Most common Zener diodes have a breakdown voltage between 5.5 and 8.5 V and they have a positive temperature drift, approximately between $+1.5$ and 5 mV/°C. As a result of its high operating voltages, though, the Zener diode is appropriately warranted in high-voltage applications with supply voltages greater than 6 to 9 V [2, 4].

Design Example 3.1: Design a 2 V zero-order reference with a 5 V $\pm$ 5% input supply voltage. The circuit should be able to drive 100 μA of load current while using only a maximum of 10 μA of quiescent current, over process variations and temperature range (0 to

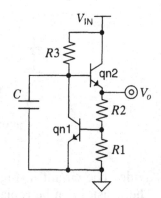

Figure 3.2   A 2 V zero-order voltage reference.

125 °C). Assume that a *vanilla* p-substrate CMOS process with an added p-base layer is to be used.

A p-substrate CMOS process with a p-base layer extension is essentially a cost-effective biCMOS process technology (i.e., a vertical NPN transistor is available where a source-drain $n+$ diffusion is used as an emitter, a p-base as a base, and an n-well as a collector). The only issue with this device is that its extrinsic saturation voltage (effective V_{SAT} is greater than the ones exhibited in standard bipolar or biCMOS technologies where a highly doped buried layer is available. This degradation results because of the parasitic resistance through the lightly doped n-well region, collector path. Other than that, the device is very useful, especially in the design of references.

Now, since a 2 V reference is desired and the supply voltage varies from 4.5 to 5.5 V, a Zener Diode cannot be used because of its high operating voltages. As a result, a forward-biased diode voltage is chosen. Figure 3.2 illustrates a circuit that fulfills the requirements stipulated in the problem statement. To generate a 2 V reference out of a nominal 0.7 V diode voltage, a base-emitter voltage (V_{BE}) multiplier is used, which is mainly composed of qn1, qn2, $R1$, and $R2$. Transistor qn2 is added to supply current without significantly affecting the output voltage. Resistor $R3$ is used only to establish current-flow through qn1 and to provide base current to qn2. Capacitor C is added for noise immunity from the positive input supply. High-frequency noise generated by the supply is shunted to ground, at the base of qn2, by capacitor C; $R3$ and C essentially form a low-pass filter with a pole at $1 \div (2\pi CR3)$. By the way, noise from ground is common mode, therefore it does not affect the performance of the circuit considerably.

The ratio of resistors $R1$ and $R2$ is adjusted to produce a 2 V output,

$$V_O = V_{BE1}\left(1 + \frac{R2}{R1}\right) \approx (0.7 \text{ V})\left(1 + \frac{R2}{R1}\right) \equiv 2\, V.$$

The absolute values of all three resistors is dependent on quiescent-current flow, which is defined to be less than 10 μA over process and temperature variations. The current through resistor R3 has more variation than $R1$ and $R2$ because its voltage changes with supply voltage. Consequently, more current is allocated to this device (i.e., 7 μA). The worst-case base-emitter voltage, with regard to current flow, occurs at opposite extremes of the temperature range (0.7 V $\pm$ (2.2 mV/°C) Δtemp). With this in mind, the resistor values are determined,

$$I_{R1} = \frac{V_{BE1}}{R1} < \frac{(0.7 \text{ V}) + (2.2 \text{ mV})(27 - 0\,°\text{C})}{R1} = \frac{0.76}{R1} \leq 3\ \mu\text{A}$$

or

$$R1 \geq 253 \text{ k}\Omega$$

and

$$I_{R3} = \frac{V_{\text{IN}} - (2 \text{ V} + V_{BE2})}{R3}$$

$$\leq \frac{5.5 \text{ V} - [2 \text{ V} + 0.7 \text{ V} - (2.2 \text{ mV})(125 - 27\,°\text{C})]}{R3} \leq 7\ \mu\text{A}$$

or

$$R3 \geq 431 \text{ k}\Omega.$$

Since integrated resistors can vary $\pm 20\%$ as a result of process variations, $R1$ and $R3$ are chosen to be 320 kΩ and 550 kΩ, respectively,

$$R1_{\text{nominal}} \equiv \frac{R1_{\text{extreme}}}{80\%} = \frac{253\,\text{k}\Omega}{0.8} = 316\,\text{k}\Omega$$

and

$$R3_{\text{nominal}} \equiv \frac{R3_{\text{extreme}}}{80\%} = \frac{431\,\text{k}\Omega}{0.8} = 539\,\text{k}\Omega.$$

Resistor $R2$, of course, is consequently designed to be 595 kΩ,

$$R2 = R1\left(\frac{V_O}{V_{BE1}} - 1\right) = (320\text{k})\left(\frac{2}{0.7} - 1\right) = 594\,\text{k}\Omega.$$

Finally, transistor qn1 is chosen to be minimum size because its effective β is typically high for the low current density required—less than 7 μA. Transistor qn2, on the other hand, is chosen to be twice the minimum size because it has to supply a total current of roughly 103 μA. The size requirements may change, of course, depending on the particular performance parameters of the process. Consequently, β versus current density curves need to be consulted. Oversizing these devices is not usually prudent since conservation of silicon area is key for maximum integration capacity. As for capacitor C, it is chosen to have a nominal pole frequency of approximately 100 kHz. Consequently, its size is designed to be 3 pF,

$$\text{Pole} = \frac{1}{2\pi CR3} \equiv 100\,\text{kHz}$$

or

$$C = \frac{1}{2\pi(100\text{k})R3} = \frac{1}{2\pi(100\text{k})(500\text{k})} = 3.2\,\text{pF}.$$

3.2   FIRST-ORDER REFERENCES

3.2.1   Forward-Biased Diode References

Forward-biased diode references with first-order or second-order compensation are commonly referred to as *bandgap references*. They are more accurate and more suitable for low-voltage operation than

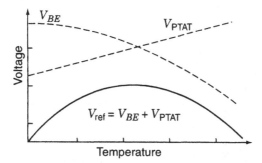

Figure 3.3   Temperature behavior of first-order bandgap references.

the Zener counterparts. Typical bandgap references have an output voltage of approximately 1.2 V, which roughly corresponds to the diode voltage at 0 °K. The basic operation relies on the temperature dependence of the base-emitter voltage of a bipolar transistor (V_{BE}) or, equivalently, a forward-biased diode. First-order bandgap voltage references compensate the linear component but ineffectively compensate the nonlinear component of the diode voltage. Summing a proportional-to-absolute temperature (PTAT) voltage and a base-emitter voltage does this type of compensation. The PTAT voltage increases linearly with temperature, thereby efficiently canceling the effect of the negative linear temperature dependence of the base-emitter voltage. The resulting reference exhibits improved temperature-variation performance over its zero-order counterpart, typically between 20 and 100 ppm/°C [5]. However, the extent of the improvement is limited by the logarithmic dependence of the diode voltage, as described equation (1.6) of Chapter 1, and whose effect is depicted by the curvature of V_{ref} in Figure 3.3. In essence, the PTAT voltage only cancels the CTAT behavior of the reference. Consequently, V_{ref} still shows the effects of the logarithmic behavior of the diode voltage.

The reference voltage is defined by first ascertaining the absolute value at room temperature (V'_{ref}). This value is not necessarily the point on the V_{ref} curve that has a slope of zero (derivative is equal to zero). Moreover, the voltage V'_{ref} is typically above the actual bandgap voltage (for a given process technology) by approximately 70 to 80 mV [1]. In the end, the absolute value of the reference will fall between 1.17 and 1.25 V. This value, of course, is typically obtained by characterizing the process for a given circuit topology. The characterization can be based either on the models available or, better yet, on experimental results obtained for the particular circuit at hand. Since

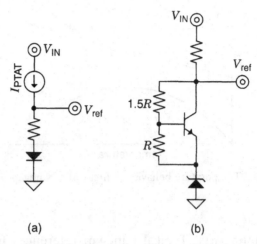

(a)   (b)

Figure 3.4   First-order voltage references.

the models are usually optimized for a wide range of operating conditions, some inaccuracies may exist when ascertaining the precise behavior under a given subset of operating conditions. In the end, unfortunately, the value of V'_{ref} is not the same for all circuits. This lack of consistency is because the effects of all the circuit components on the reference are dictated by the topology itself as well as by its biasing condition. For instance, the TC of the resistors used in the circuit (for the PTAT current generator) affects the absolute value of the base-emitter voltage and the collector current flow at a given temperature. These and other effects will be discussed in greater detail in Chapter 4.

Once V'_{ref} is obtained, the value of the base-emitter voltage at room temperature is subtracted from V'_{ref}; thus, a value for the PTAT voltage can be calculated. Figure 3.4(a) shows a simple topology that achieves the objectives described above and is compatible with CMOS and bipolar processes. It is noteworthy to mention that the resistors used in generating the PTAT current should match the resistor in the figure at hand to maintain the quality of the PTAT voltage. The best value for V'_{ref} occurs when the voltages of V_{ref} at the extremes of the temperature range are roughly equal, as depicted by Figure 3.3. Actual production material, because of process gradients and variants, will not have these two values equal for all parts, for all wafers, and for all lots. However, the overall shape of the curve should be approximately the same after the parts have been trimmed. The effects of Early voltage (or channel-length modulation) and package shifts have been neglected thus far. The former effect is particular to each design and can be minimized through careful device choice and placement (i.e., orientation and relative position of devices). The package shift is an effect that should be either trimmed after the fact, modeled and taken into account before the fact, or simply superimposed onto the overall variation performance of the part. The particulars of package shifts are addressed later in Chapter 4.

3.2.2 Zener References

The temperature-drift performance of a Zener diode can be significantly improved by cascading one or several elements with negative temperature coefficients (TCs), i.e., forward-biased diodes (TC $\approx$ -2.2 mV/°C per diode). This performance enhancement, however, comes at the expense of increased output voltage. For instance, if the temperature drift of the Zener diode is assumed to be $+5.5$ mV/°C, then approximately 2.5 forward-biased diode voltages (2.5×-2.2 mV/°C $= -5.5$ mV/°C) are needed to cancel the linear TC. Consequently, the output voltage of the reference becomes roughly 5.5 V + (2.5 × 0.7) V or 7.25 V, assuming the Zener voltage is around 5.5 V. One embodiment of this concept is illustrated in Figure 3.4(b). The resistor connected to input voltage V_{in} can be replaced with any current source or even a JFET. The other two resistors comprise a diode voltage multiplier, i.e., (V_{BE} [1 + 1.5R/R] = 2.5V_{BE}). In a purely CMOS environment, the bipolar transistor and the corresponding resistors would have to be replaced with two or three p-n junction diodes in series. This modification would degrade the temperature-drift performance since only an integer number of diode voltages is feasible. The integer limitation is circumvented if a standard voltage multiplier circuit is designed. This design change, however, increases the complexity of the overall circuit.

Design Example 3.2: Design a first-order bandgap reference for a temperature ranging from 0 to 125 °C, assuming that the bandgap voltage (V_{go}) is 1.2 V and the process constant η is approximately 3.6.

The circuit is illustrated in Figure 3.5. The performance of this circuit is independent of the startup current, which is of the continuous conduction type. Startup current $I_{startup}$, however, does not affect transistors qn2 or qn3, which are the critical transistors of the reference circuit. In fact, the sum of the currents flowing through mp4 and

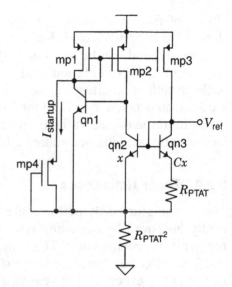

Figure 3.5 First-order bandgap reference.

qn1 is exactly equal to the PTAT current (current flowing through qn2 and qn3), assuming there are no mismatch errors.

The first portion of the circuit to be designed is the PTAT generator block. This block is comprised of mp1 through mp3, qn1 through qn3, and R_{PTAT}. Transistor qn2 is chosen to have an area equal to four times larger than the minimum possible size for an NPN device. A nonminimum device size is chosen to maximize the matching capabilities between transistors qn2 and qn3. Current density curves (β curves) should also be taken into account but will not usually be a limiting factor. Now, the current through qn3 is chosen to be 15 μA. Lower currents can be chosen; however, low-current cells are more vulnerable to noise. The size of qn3 needs to be larger than qn2 and, for this particular design, it is chosen to be 8 times larger than qn2. At this point, resistor R_{PTAT} is simply

$$R_{PTAT} = \frac{V_T \ln C}{I_{PTAT}} = \frac{(25.8\text{m}) \ln 8}{15 \ \mu\text{A}} \approx 3{,}580 \ \Omega.$$

Transistor qn1 is only a feedback device and therefore can be chosen to be minimum size. Transistors mp1 through mp3 compose a mirror, which is chosen to source equal currents. Their size is chosen to maximize accuracy (i.e., $W/L = 100/25$). A long channel length is chosen to minimize channel-length modulation errors and a long width is chosen to maximize matching performance. The saturation voltage ($V_{GS} - V_t$ or V_{GSt}) is chosen to ensure that it does not operate in subthreshold (i.e., $V_{GSt} > 150$ mV). The matching capability for devices operating in subthreshold is not as good as in strong inversion. Transistor mp4 is used as a startup device pulling some current from device mp1. Its current must not exceed, or even approach, the PTAT current; otherwise, transistor qn1 could be starved of current. Thus, a current of approximately 1 μA is chosen,

$$V_{SGt4}V_{SG4} - |V_t| = \left(5 \text{ V} - V_{SG1} - V_{R_{PTAT2}}\right) - |V_t|$$

$$= 5 \text{ V} - V_{SGt1} - 0.6 \text{ V} - 2|V_t|$$

$$\approx 3 \text{ V} - \sqrt{\frac{2(15 \ \mu A)}{15\mu(100/25)}} \approx 2.3 \text{ V}$$

$$\equiv \sqrt{\frac{2(1 \ \mu A)}{15\mu(W/L)_{mp4}}}$$

or

$$(W/L)_{mp4} = \frac{1}{40},$$

where the transconductance parameter K' is assumed to be 15 $\mu A/V^2$ and the threshold voltage $|V_{tp}|$ is assumed to be 0.7 V. The voltage across resistor R_{PTAT2} ($V_{R_{PTAT2}}$) is momentarily chosen to be 0.6 V at room temperature. This assumption is reasonable for the above derivation since the ultimate reference voltage will be close to 1.2 V and V_{BE} is roughly between 0.6 V and 0.7 V at room temperature. The error that arises from this conjecture will not have significant impact on the overall performance of the circuit since $I_{\text{start-up}}$ can exhibit significant variation before I_{PTAT} is affected. The current through qn1 is simply the difference between the PTAT current (15 μA) and the startup current (1 μA) (i.e., approximately 14 μA at room temperature).

The reference voltage is equal to the sum of base-emitter voltage V_{BE2} and the PTAT voltage across resistor R_{PTAT2}. The way to design the value of this resistor is to make a first-order approximation of the base-emitter voltage and to determine the PTAT voltage required to produce a zero TC. The approximation will neglect the higher-order nonlinear components of the base-emitter voltage,

$$V_{BE2} \approx \left[V_{go} + (\eta - 1)V_{T_r}\right]$$

$$-\left[V_{go} - V_{BE2}(T_r) + (\eta - 1)V_{T_r}\right]\frac{T}{T_r},$$

where V_{T_r} is the thermal voltage at room temperature T_r and 1 is substituted for the constant x in equation (1.7) of Chapter 1 since the collector current is PTAT ($I_{C2} \propto T^1$). For the voltage reference to have zero TC, its first derivative must be equal to zero, i.e.,

$$\frac{\partial V_{\text{ref}}}{\partial T}\bigg|_{T=T_r} = \frac{\partial(V_{BE2} + V_{\text{PTAT}})}{\partial T}\bigg|_{T=T_r}$$

$$= -\left[\frac{V_{go} - V_{BE2}(T_r) + (\eta - 1)V_{T_r}}{T_r}\right]$$

$$+ \left[3\frac{I_{\text{PTAT}}(T_r)}{T_r}\right]R_{\text{PTAT2}} \equiv 0,$$

where V_{PTAT} is the PTAT voltage across resistor R_{PTAT2}. The multiplier "3" in the above relation comes from the fact that the current through R_{PTAT2} is equal to $3I_{\text{PTAT}}$. Consequently, the value for R_{PTAT2} can be found to be 13,300 Ω,

$$R_{\text{PTAT2}} = \frac{V_{go} - V_{BE2}(T_r) + (\eta - 1)V_{T_r}}{3I_{\text{PTAT}}(T_r)}$$

$$= \frac{1.2 \text{ V} - 0.67 \text{ V} + 67.1 \text{ mV}}{3(15 \text{ }\mu\text{A})} \approx 13,300 \text{ }\Omega,$$

where V_{BE2} is assumed to be 0.67 V at room temperature. The resulting reference voltage at room temperature is

$$V_{\text{ref}}(T_r) = V_{BE}(T_r) + V_{\text{PTAT}}(T_r) = 0.67 \text{ V} + 3(15 \text{ }\mu\text{A})(13,300 \text{ }\Omega)$$

$$\approx 1.268 \text{ V}.$$

At this point, the circuit is simulated to determine its temperature-drift performance. Most likely, the performance will not be at its optimum point. One of the reasons for this deviation is that the logarithmic term was neglected. In other words, the PTAT voltage may need some adjustment to center the higher-order components of the base-emitter voltage in the given temperature range. Furthermore, some of the parasitic errors of the circuit were not taken into account (i.e., the channel-length modulation error of the PMOS transistors, the temperature drift of the resistors, the Early-voltage effect of the NPN transistors, etc.). However, the simulator is used to obtain the best value of R_{PTAT2} for which the reference voltage will exhibit its lowest possible temperature variation.

3.3   SECOND-ORDER REFERENCES (CURVATURE CORRECTION)

Low dynamic range and low-voltage swings are natural circuit characteristics that result because of the growing market demand for battery-operated electronics. A battery-powered IC intrinsically requires low quiescent current flow and low-voltage operation. The signal-to-noise ratio (SNR) is relatively low because noise does not scale with decreasing supply voltages. Unfortunately, SNR is further degraded by an inherent increase in broad-spectrum noise resulting from low quiescent-current flow. These characteristics force increasingly stringent specification requirements on integrated references. Typical first-order bandgap references are no longer adequate for many high-performance systems. Consequently, higher-order references—curvature-corrected references—are necessary!

In addition to canceling first-order terms, curvature-corrected bandgap references attempt to approximately cancel the nonlinear component of the base-emitter (diode) voltage. Essentially, the second-order term in the Taylor-series expansion of the diode relationship is canceled, leaving only third-order and higher-order components. The classical method for doing such compensation is through the addition of a squared PTAT term ($PTAT^2$), a quadratic component, to the output voltage relation of first-order bandgap references [6]. The idea is to offset the negative temperature dependence of the logarithmic term in V_{BE} with a positive parabolic term. Figure 3.6 shows the typical temperature-drift performance achieved

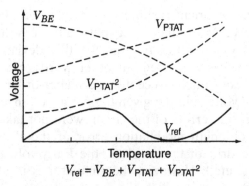

$$V_{ref} = V_{BE} + V_{PTAT} + V_{PTAT}^2$$

Figure 3.6 Squared PTAT curvature-correction method for bandgap references.

by PTAT^2 curvature-corrected bandgap references. The lower temperature range is predominantly controlled by the base-emitter voltage and the linear PTAT component ($V_{BE} + V_{PTAT}$). As a result, the first half of the temperature range exhibits the curvature of a first-order bandgap reference. The squared PTAT voltage term (V_{PTAT^2}), however, becomes considerably large as the temperature increases. This behavior is used to cancel the increasingly negative temperature dependence of the base-emitter voltage at higher temperatures. Consequently, the curvature depicted by V_{ref} in Figure 3.6 results. Prevailing curvature-corrected bandgap references achieve a temperature-drift performance of roughly 1 to 20 ppm/°C [5]!

Generally, the base-emitter (diode) voltage is described by

$$V_{BE}(T) = V_{go} - BT^1 - Cf(T), \qquad (3.1)$$

where V_{go} (extrapolated diode voltage at 0 °K), B, and C are temperature-independent constants and T refers to temperature, which is more explicitly illustrated in equation (1.6) of Chapter 1. First-order references sum a positive linear temperature-dependent voltage to a base-emitter voltage to cancel the effects of constant B. The goal of higher-order bandgap references, on the other hand, is to sum a temperature-dependent voltage exhibiting both a positive linear and a positive nonlinear temperature dependence, i.e.,

$$V_{ref} = V_{BE}(T) + V_x(T) = V_{BE}(T) + DT^1 + Ef_2(T) \approx V_{go}, \qquad (3.2)$$

where voltage V_x, which is equal to $DT^1 + Ef_2(T)$, is generated by the reference circuit and constant D and $Ef_2(T)$ approximately equal B and $Cf(T)$, respectively. The most common technique used to generate $Ef_2(T)$ is the squared PTAT compensation scheme, as mentioned above. In this method, a squared PTAT dependent voltage is summed with a PTAT and a diode (base-emitter) voltage.

The process of designing a curvature-corrected bandgap circuit is inherently iterative in nature. This characteristic exists because there usually is more than one combination of term-coefficients that produces similar temperature-variation performance. The procedure is partially automated by determining the coefficients through the use of a spreadsheet. Practical circuits, however, also suffer from such effects as Early voltage (channel-length modulation), bulk effects, parasitic current paths, finite betas (β), and so on. As a result, for each combination of coefficients, the parasitic effects may be different. In other words, the process of ascertaining the voltage at which the reference will have its best temperature-drift performance is not straightforward. The simulator helps project where it lies but the prediction is as good as its models. Simulations of the specific circuit, however, can give the designer a good indication as to where the reference is best centered for the particular process. In determining the absolute value of the reference voltage, the design should also take into account initial accuracy by including a trim network and by relying on models and experimental measurements for the ratio of the coefficients of the temperature components. Sometimes, experimental measurements of the actual circuit are necessary to effectively infer the voltage where the reference is best centered. These measurements become intrinsic when the package itself imposes significant offset and temperature drift on the circuit—a factor that is not typically modeled in simulations.

Figure 3.7(a) illustrates the realization of second-order correction on a distinctively popular bandgap circuit, the Brokaw bandgap reference, which was partially used in Design Example 3.2. In this instance, the PTAT current generator and the bandgap circuit are integrated into one cell. The loop composed of qn1, qn2, and R produces the PTAT current that flows through resistors $R1$ and $R2$. If the squared PTAT current is disregarded, the resulting circuit is a first-order bandgap reference. The reference is simply the sum of the base-emitter voltage of qn1 and the PTAT voltage imposed across $R1$ and $R2$. Second-order correction is finalized when the squared PTAT current is forced to flow through $R2$. Consequently, the relation that governs

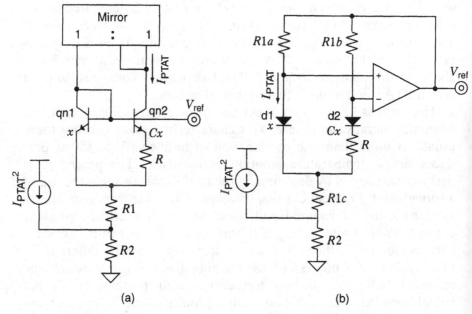

Figure 3.7 Second-order curvature-corrected bandgap references.

the reference voltage is described by

$$V_{ref} = V_{BE1} + 2I_{PTAT}(R1 + R2) + I_{PTAT^2}R2, \qquad (3.3)$$

where V_{BE1} is the base-emitter voltage of qn1, I_{PTAT} is the PTAT current, and I_{PTAT^2} is the squared PTAT current. The mirror, as is the case for most topologies, can be replaced with any configuration that ensures that the collector currents of qn1 and qn2 are equal. Figure 3.7(b) shows an equivalent CMOS realization where an operational amplifier configuration is utilized to ensure that the PTAT currents through diodes d1 and d2 are equal. The relationship of the reference voltage for this circuit is similar to that of the previous one except that the coefficient of the PTAT term is slightly different,

$$V_{ref} = V_{D1} + 2I_{PTAT}\left(\frac{R1a}{2} + R1c + R2\right) + I_{PTAT^2}R2, \quad (3.4)$$

where V_{D1} is the diode voltage across diode d1. It is noted that $R1a$ and $R1b$ are equal and should be laid out to match well!

3.4 STATE-OF-THE-ART CURVATURE-CORRECTION TECHNIQUES

The squared PTAT curvature-correction scheme, which yields second-order references, is not the only technique used to cancel the higher-order components of the diode voltage. Other design approaches, in fact, can be adopted to effectively perform curvature correction, yielding both second-order and higher-order circuits. Some of these other methods will be explored in the following subsections, ranging from a temperature-dependent resistor ratio technique and a β-dependent voltage approach to ultimately exacting cancellation techniques. These, of course, will be analyzed and compared for performance and versatility.

3.4.1 Temperature-Dependent Resistor Ratio

A nonlinear component can be generated by exploiting the temperature dependence of different resistors in a given process technology. Typically, circuit designers attempt to mitigate the effects of the resistors' temperature coefficient (TC) since they are considered parasitic to the design. Figure 3.8, however, illustrates a technique where the TC of different resistors is used to compensate the higher-order components present in a first-order bandgap cell. The goal is to create a squared PTAT voltage without significantly increasing overhead. Typical implementations of the squared PTAT voltage are more complex and require more quiescent-current flow. The reference voltage of this sample circuit is

$$V_{\text{ref}} = \frac{AV_{BE}}{R_1}R_4 + \frac{BV_T}{R_2}R_4 + \frac{BV_T}{R_2}R_3, \tag{3.5}$$

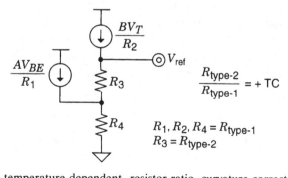

Figure 3.8 A temperature-dependent, resistor-ratio, curvature-corrected reference.

where A and B are constants, V_T is the thermal voltage, and all the resistors are constructed using the same type of material, except for resistor R_3. The ratio of resistors R_3 and R_2 must be roughly linearly positive to generate the desired squared PTAT dependence. Toward this end, a survey of the process technology is necessary to ascertain if there are any resistor-candidates that match the identified criteria. For instance, a low TC resistor can be used for all the resistors except for R_3, which should have a relatively large positive TC. Consequently, the first two components of the reference voltage relation achieve first-order cancellation while the third term introduces curvature correction.

The advantages of this technique over typical squared PTAT realizations are reduced overhead and simplicity. Only one additional resistor is necessary to convert a first-order bandgap reference into a second-order circuit. There is no significant cost, if any at all, incurred in quiescent-current flow or silicon area. The circuit, however, suffers from a subtle disadvantage; namely, the large mismatch between the two types of resistors used must be taken into account in the design of the trim network. This means that more bits of trim may be required. The circuit also exhibits a more significant disadvantage in that the design is highly dependent on the process. The process technology must have two types of resistors whose ratio yields a linearly positive TC. In the end, the design of the squared PTAT coefficient varies substantially from process to process due to inherent differences in technology; however, the circuit realization is simple and cost effective.

3.4.2 Diode Loop

The curvature-correcting component of a high-order bandgap reference can also be effectively generated through the use of different temperature-dependent currents and a diode-voltage loop [7]. The circuit realization of this particular technique is illustrated in Figure 3.9. Nonlinear current component I_{NL} is defined by different temperature-dependent currents and by a transistor loop comprised of qn1, qn2, and R_3,

$$I_{NL} = \frac{V_T}{R_3} \ln\left(\frac{I_{C1}A_2}{A_1 I_{C2}}\right) = \frac{V_T}{R_3} \ln\left(\frac{2I_{\text{PTAT}}}{I_{NL} + I_{\text{constant}}}\right), \qquad (3.6)$$

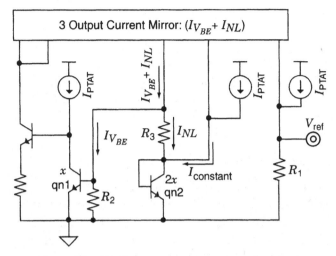

Figure 3.9   Diode-loop curvature-corrected method.

where $I_{constant}$ is a current whose temperature dependence is dominated only by the temperature coefficient of the resistors used in the circuit. The nonlinear temperature dependence of voltage $I_{NL}R_3(V_{NL})$ is designed to cancel the effects of the higher-order components introduced by the base-emitter voltage of qn1,

$$
I_{constant} = I_{PTAT} + I_{V_{BE}} + I_{NL} \approx \frac{V_T}{R_x} + \frac{V_{BE1}}{R_2} + \frac{V_{NL}}{R_3} \approx \frac{V_{go}}{R_2}, \quad (3.7)
$$

where I_{C1} (I_{C2}) and A_1 (A_2) are the collector current and area of qn1 (qn2), respectively, V_T is the thermal voltage, and R_x and R_3 are chosen appropriately. Current I_{NL} is a logarithmic function of itself, therefore it exhibits a nonlinear behavior. The resulting reference voltage for this circuit (V_{ref}) is

$$
V_{ref} = I_{constant}R_1 \approx \frac{R_1}{R_2}V_{go}. \quad (3.8)
$$

This circuit is suitable for low-supply-voltage applications. The headroom limitation is ultimately defined by a diode-connected device, a drain-source (collector-emitter) voltage, and a relatively small resistive voltage drop, which results in a voltage headroom limit roughly between 0.9 and 1.1 V, in a typical process. The technique, however, lacks the *elegance* of a simple solution. The circuit presented has inherently more quiescent-current flow than the previously surveyed squared PTAT scheme; in other words, this circuit has more current-sensitive paths to ground. A physical realization of this particular reference achieved a TC of approximately ±3.0 ppm/°C with a total quiescent current flow of 95 μA [7].

3.4.3 β Compensation

The exponential temperature dependence of the forward-current gain (β) of NPN transistors can be exploited to correct the nonlinear behavior of the diode voltage, $\beta \propto e^{-1/T}$ [8]. Basically, current gain β increases exponentially with rising temperatures, which is equivalent to saying that bipolar devices become stronger with ascending temperatures. Figure 3.10 illustrates one simple circuit realization that capitalizes on this exponential characteristic. The circuit generates a negative reference with respect to the ground potential. Its translation to a circuit architecture with a positive output voltage is readily achieved by designing the complement of the structure shown (i.e., PNP device version), or by modifying another first-order bandgap reference to adopt this technique. The resulting reference voltage for this circuit is

$$V_{\text{ref}} = -\left(AT + \frac{BT}{\beta}\right)R - V_{BE}, \tag{3.9}$$

where A and B are temperature-independent constants and T is temperature. The current sources (or current sinks) fall under the proportional-to-absolute temperature (PTAT) category; thus, they can

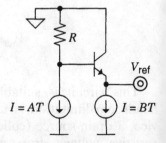

Figure 3.10 Curvature-corrected bandgap reference using a β-compensating scheme.

be represented with a linear first-order relationship. This curvature-correcting scheme reaps the benefits of being simple and achieving good overall performance with low quiescent currents. The lower limit of the power supply voltage (difference between extreme supplies) is approximately 1.5 V. This limit is derived from the sum of the reference voltage and the voltage across the current sources, which is essentially a drain-source (collector-emitter) voltage drop.

3.4.4 Piecewise-Linear Current-Mode Technique

The basic theory behind the curvature-correcting component of a bandgap reference is that it must increase with temperature in a nonlinear fashion. In other words, its dependence to temperature must have equal or higher than second-order components. For the case of the squared PTAT technique, the nonlinear component has a quadratic dependence to temperature—second-order component. The nonlinear current, though, can be alternatively exponential or even piecewise linear. The circuit in Figure 3.11(a) exemplifies a method by which nonlinear current I_{NL} *is generated* [9]. *Current* I_{NL} *is used as the curvature-correcting component of a high-order bandgap reference.*

The circuit generates current I_{NL} through a current-mode operation. Its resulting dependence to temperature is piecewise linear. The technique is based on the nodal subtraction of currents and on the

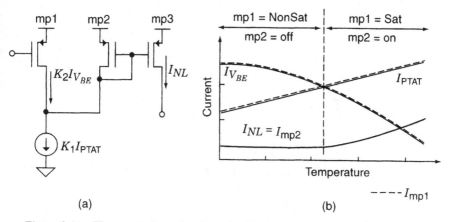

(a)   (b)

Figure 3.11   The generation of a piecewise-linear curvature-correcting current.

characteristics of nonideal transistors. Figure 3.11(b) graphically illustrates the operation of the circuit throughout the operating temperature range. Transistor mp1 acts like a nonideal source of current that is proportional to a base-emitter voltage, equivalent to a first-order complementary-to-absolute temperature (CTAT) dependence. For the lower half of the temperature range, PTAT current I_{PTAT} is less than the supplied V_{BE}-dependent current ($I_{V_{BE}}$) if device mp1 were to operate in the saturation region. However, mp1 actually operates in the linear region supplying only I_{PTAT}; thus, mp2 does not conduct any current. For the upper half of the temperature range, I_{PTAT} becomes larger than the theoretical value of $I_{V_{BE}}$. Consequently, mp1 becomes saturated and supplies current $I_{V_{BE}}$, thereby forcing mp2 to source the difference. The resulting current through transistor mp3 is nonlinear—practically zero during the first half of the temperature range and nonlinearly increasing throughout the latter half. This relationship is described by

$$
I_{NL} = \begin{cases} 0 & I_{V_{BE}} \geq I_{PTAT} \\ K_1 I_{PTAT} - K_2 I_{V_{BE}} & I_{V_{BE}} < I_{PTAT} \end{cases}, \qquad (3.10)
$$

where K_1 and K_2 are constants defined by the ratios of the mirroring transistors defining I_{PTAT} and $I_{V_{BE}}$.

Ultimately, curvature correction is achieved by combining the three temperature-dependent elements of Figure 3.11(b) to yield an output voltage with reduced temperature-drift variation. From the practical design standpoint, the temperature range is partitioned in two: (1) the

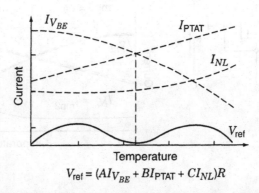

$$V_{ref} = (AI_{V_{BE}} + BI_{PTAT} + CI_{NL})R$$

Figure 3.12 Temperature dependence of the piecewise-linear curvature-corrected bandgap reference.

range for which the nonlinear current is zero and (2) the range for which the nonlinear current is nonzero. As a result, the reference voltage V_{ref} can be compensated to exhibit the behavior that is graphically described in Figure 3.12. The reference voltage at the lower half of the temperature range behaves essentially like a first-order bandgap since the nonlinear component I_{NL} is zero. At higher temperatures, the behavior is similar to that of the lower temperatures but the operation is not. The nonlinear dependence of $I_{NL}(K_1 I_{PTAT} - K_2 I_{V_{BE}})$ is designed to diminish the nonlinear effects of the base-emitter voltage. Consequently, the addition of currents $AI_{V_{BE}}$, BI_{PTAT}, and CI_{NL} at the upper temperature range generates a curvature-corrected trace whose behavior is depicted by V_{ref} in Figure 3.12. This figure assumes that the final implementation uses a current-mode output stage where temperature-dependent currents are summed to flow through a resistor. Needless to say, there are other ways of introducing this piecewise-linear current to a bandgap reference circuit.

3.4.5 Matched-Nonlinear Correction

Up to this point, the approach to curvature correction has been to "compensate" and not to "cancel" the nonlinearity of the diode voltage. In other words, the correcting component does not exhibit the same dependence to temperature as the nonlinear term of the diode voltage relationship. Diode voltage V_D exhibits the following relationship:

$$V_D \approx V_{go} - \frac{T}{T_r}\left[V_{go} - V_D(T_r)\right] - (\eta - x)V_T \ln\left(\frac{T}{T_r}\right), \quad (3.11)$$

where V_{go} is the extrapolated bandgap voltage at 0 °K, T is temperature, $V_D(T_r)$ is the diode voltage at room temperature T_r, η is a temperature-independent and process-dependent constant ranging from 3.6 to 4, x refers to the temperature dependence of the current forced through the diode ($I_D = DT^x$, where D is a constant and x equals 1 for a PTAT current), and V_T is the thermal voltage. Thus far, the correction terms explored have not had the logarithmic relationship required to completely cancel the third term of the diode voltage relationship.

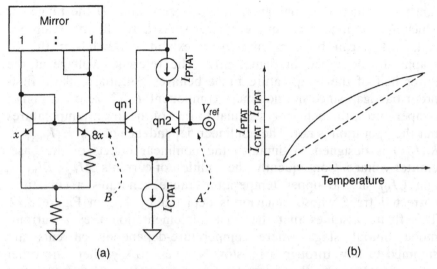

(a)   (b)

Figure 3.13   Matched-nonlinear curvature-corrected bandgap reference.

Figure 3.13 illustrates an approach where an attempt is made to more accurately *cancel* the aforementioned nonlinear dependence. This circuit is a simple and practical embodiment of the concepts introduced in [10]. The goal is to generate a voltage that has the logarithmic temperature dependence of the diode voltage ($T \ln T$). Intuitively, if the ratio of the currents flowing through a couple of diodes were to be PTAT, then the voltage difference (ΔV_{BE}) of the devices is proportional to $T \ln T$. Transistors qn1 and qn2 in Figure 3.13 exemplify this condition,

$$V_{AB} = V_{BE2} - V_{BE1} = V_T \ln\left(\frac{I_{PTAT}}{I_{CTAT} - I_{PTAT}}\right)$$

$$\approx V_T \ln\left[\frac{K_1 T}{K_2 - (K_1 + K_3)T}\right], \tag{3.12}$$

where V_{AB} is the voltage difference between nodes "A" and "B", I_{PTAT} has a linearly positive TC defined by temperature-independent constant K_1, and I_{CTAT} has a linearly negative TC defined by temperature-independent constants K_2 and K_3 ($I_{CTAT} = K_2 - K_3 T$). As a result, the term within the logarithm has a positive concave relationship with respect to temperature, as illustrated in Figure 3.13(b). The premise for this behavior is, of course, that I_{CTAT} is greater than

I_{PTAT} throughout the temperature range and that the junction area of qn1 is greater than or equal to that of device qn2. Additionally, transistors qn1 and qn2 must match well. Resulting term V_{AB}, though not exactly equal, does exhibit a similar logarithmic dependence to the nonlinear term of the diode relationship. As a result, the addition of voltage V_{AB} to the first-order reference voltage V_B converts the circuit shown in Figure 3.13(a) into a curvature-corrected bandgap reference. The shape of the curve of this higher-order reference is equal to that of a first-order bandgap reference (concave with a single inflexion point) except that the total temperature-drift variation is potentially less than 10 ppm/°C [10].

Multiple V_{AB} voltages may be summed to the output to optimize the nonlinear cancellation. This incremental summation would ease the burden of generating a relatively large ΔV_{BE} voltage in a single V_{AB} cell. Incidentally, the temperature dependence of the term contained within the logarithm can also be manipulated to more closely resemble the logarithmic component of the diode voltage. This achieves an even better approximation of the $T \ln T$ component. For instance, a $T \ln T$ term is generated if the collector current of transistor qn1 in Figure 3.13 were to be constant. This characteristic is achieved by forcing the tail current of the ΔV_{BE} cell (presently I_{CTAT}) to sink $I_{\text{PTAT}} + I_{\text{constant}}$. A first-order temperature-independent current (I_{constant}) is designed by properly summing I_{PTAT} and I_{CTAT}. Consequently, the new tail current is equal to $K_4 I_{\text{PTAT}} + I_{\text{CTAT}}$, where K_4 is independent of temperature. Constant K_4 is designed such that the total tail current is equal to the PTAT current flowing through transistor qn2 in addition to a first-order temperature-independent current.

The circuit illustrated in Figure 3.14 is another realization of the matched $T \ln T$ concept for curvature-corrected bandgap references. This version is a variation of the one proposed by [11]. The loop comprised of resistor R_x and transistors qn1 and qn2 generates the $T \ln T$ dependence. The current flowing through resistor R_x is

$$I_x = \frac{V_T}{R_x} \ln\left(\frac{BI_{\text{PTAT}}}{I_{\text{PTAT}} + I_{\text{CTAT}}} \right) \approx \frac{V_T}{R_x} \ln\left(\frac{BI_{\text{PTAT}}}{I_{\text{constant}}} \right) \propto T \ln(AT), \quad (3.13)$$

where A and B are temperature-independent constants and I_{constant} is a first-order temperature-independent current. Current I_x is summed with the PTAT current, which flows through resistor R, thus effectively adding the matched $T \ln T$ component to the reference voltage

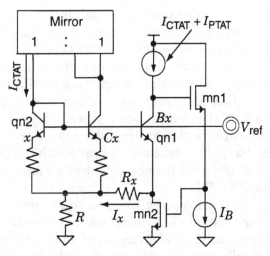

Figure 3.14 Version 2 of the matched-nonlinear curvature-corrected bandgap reference.

relationship,

$$V_{\text{ref}} = V_{BE2} + 2I_{\text{PTAT}} R + I_x R$$

$$= V_{BE2} + 2I_{\text{PTAT}} R + \frac{RV_T}{R_x} \ln\left(\frac{BI_{\text{PTAT}}}{I_{\text{constant}}}\right). \quad (3.14)$$

Consequently, the following design relationship must be obeyed to effectively cancel the logarithmic component of the diode voltage:

$$\frac{RV_T}{R_x} \ln\left(\frac{BI_{\text{PTAT}}}{I_{\text{constant}}}\right) \equiv (\eta - 1)V_T \ln\left(\frac{T}{T_r}\right)$$

or

$$R_x \equiv \frac{R}{(\eta - 1)} \ln\left(\frac{BI_{\text{PTAT}}T_r}{I_{\text{constant}}T}\right). \quad (3.15)$$

Since I_{PTAT} is directly proportional to temperature, temperature T contained within the logarithm is canceled. Current source I_B and transistors mn1 and mn2 comprise a negative feedback loop used to regulate current I_x. Unlike the previous example, this circuit does not need multiple cells to obtain the proper coefficient for the logarithmic term. The coefficient, for this instance, is controlled by the ratio of resistors R and R_x, which has considerable flexibility. The number of additional devices required to convert a first-order reference to a matched-order version is relatively low. As a result, the additional overhead of quiescent current is also low.

3.4.6 Exact Method

Ideally, the correcting-nonlinear term of a high-order bandgap reference is "exactly" equal and opposite to the logarithmic component of the diode voltage. The matched technique approaches it by attempting to mathematically mimic the nonlinearity. However, the cancellation is theoretically enhanced if the correcting-nonlinear term is itself derived from a diode voltage. Consequently, the correcting term would have all the same higher-order temperature-dependent components that the diode voltage has. Figure 3.15 illustrates a circuit that strives to innately cancel the diodes nonlinearity by manipulating the variable x within the diode voltage relationship (equation (3.11)) to equal the extrapolated process-dependent constant η, which is approximately 4 [12]. The variable x is dependent on the collector current. When the collector current for an NPN device is PTAT ($I_C \propto T^1$), x is equal to 1. On the other hand, x is equal to zero when the collector current is independent of temperature ($I_C \propto T^0$). This dependence to the temperature behavior of the collector current is exploited by the NPN diode loop shown in Figure 3.15. Temperature-independent current

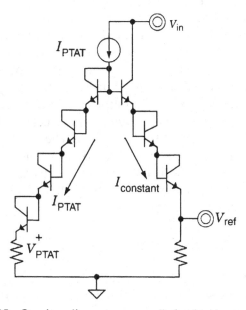

Figure 3.15 Quasi-nonlinear term cancellation bandgap reference.

I_{constant} is derived from the reference voltage itself. The PTAT current flows through four diodes while the constant current flows through three. The resulting reference voltage is described by

$$V_{\text{ref}} = V_{\text{PTAT}} + 4V_{BE\text{-PTAT}} - 3V_{BE\text{-constant}}, \qquad (3.16)$$

where $V_{BE\text{-PTAT}}$ and $V_{BE\text{-constant}}$ are base-emitter voltages with I_{PTAT} and I_{constant} as collector currents, respectively. Furthermore, both current densities must be equal at the reference temperature T_r ($V_{BE\text{-PTAT}}(T_r) - V_{BE\text{-constant}}(T_r) = 0$). If the diode expression of equation (3.11) is substituted into this last equation, the reference voltage becomes

$$V_{\text{ref}} = V_{\text{PTAT}} + V_{go} - \frac{T}{T_r}\left[V_{go} - V_{BE}(T_r)\right] - (\eta - 4)V_T \ln\left(\frac{T}{T_r}\right);$$

$$(3.17)$$

however, term ($\eta - 4$) approximately cancels to zero since η is roughly 4! The remaining terms are only first order with respect to temperature and therefore can be designed to cancel each other, yielding a theoretically temperature-independent reference voltage.

The performance of this circuit is limited by several factors: process-dependent constant η is, for the most part, a noninteger value, the power supply voltage is limited to be greater than roughly 4 V, and the TC of the output resistor must be low. If the value of η is not an integer, then exact cancellation of the logarithmic dependence of the diode voltage cannot be achieved. Furthermore, I_{constant} ceases to be independent of temperature if an output resistor with significant TC is used. Nevertheless, the circuit provides an accurate reference voltage with potentially low quiescent-current flow. The most important drawback, however, is that the circuit topology is not appropriate for low-voltage applications, which are currently driving a large portion of the market demand. This disadvantage results because of the stack of multiple base-emitter voltages,

$$V_{\text{in}} \geq V_{\text{PTAT}} + 4V_{BE} + V_{\text{current-source}} \approx 4\text{--}5 \text{ V}, \qquad (3.18)$$

where V_{in} is the input power supply voltage and $V_{\text{current-source}}$ is the voltage overhead associated with the current source (source-drain or emitter-collector voltage).

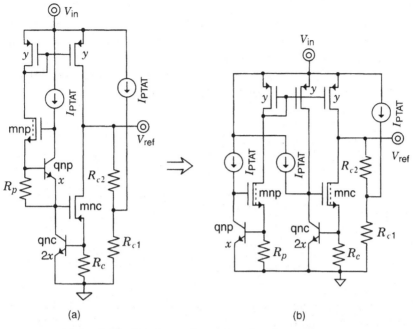

Figure 3.16   Quasi-exact low-voltage bandgap reference.

Figure 3.16 illustrates an exact reference that circumvents the limitations of η and supply voltage, which were prevalent in the previous circuit. The input voltage must exceed 4 V and η must be an integer value for the circuit in Figure 3.15 to operate properly. On the other hand, the minimum supply is lower and η can be any noninteger value in the circuits shown in Figure 3.16. Transistors mnp and mnc are natural NMOS devices; natural n-type MOS transistors exhibit a typical threshold voltage of 0 V. They do not have to be natural devices but they are chosen as such to exploit their low voltage characteristics. These transistors sense the current flowing through the resistors tied across the base and emitter of qnp and qnc. Consequently, applying Kirchhoffs Current Law (KCL) on the output derives the reference voltage,

$$V_{\text{ref}} = I_{\text{PTAT}} R_{o1} + \left(\frac{V_{BEp}}{R_p} - \frac{V_{BEc}}{R_c} \right) (R_{o1} + R_{o2}), \qquad (3.19)$$

where V_{BEp} and V_{BEc} are the base-emitter voltages of qnp and qnc, respectively. The current flowing into the collector of qnc is the summation of I_{PTAT} and V_{BEp}/R_p, which is designed to yield a first-order temperature-independent current. As a result, substituting equation (3.11) into (3.19) while taking into account that qnc's collector current is PTAT ($x = 1$ in equation (2.11)) and qncs collector current is temperature-independent ($x = 0$ in equation (2.11)) yields the following expression for the reference voltage:

$$V_{ref} = I_{PTAT}R_{o1} + \left(\frac{1}{R_p} - \frac{1}{R_c} \right)(R_{o1} + R_{o2})$$

$$\times \left(V_{go} - \frac{T}{T_r}\left[V_{go} - V_{BEp}(T_r) \right] \right)$$

$$-\left[\frac{(\eta - 1)}{R_p} - \frac{\eta}{R_c} \right](R_{o1} + R_{o2})V_T \ln\left(\frac{T}{T_r} \right), \quad (3.20)$$

where V_{BEp} is designed to equal V_{BEc} at reference temperature T_r (same current density at T_r: $I_{PTAT} \div x|_{T_r} \approx (I_{PTAT} + V_{BEp}/R_p) \div 2x|_{T_r}$). To completely eliminate the temperature dependence of V_{ref}, $I_{PTAT}R_{o1}$ is designed to cancel the linear temperature component of the base-emitter voltage while the following design relation is formulated to cancel the nonlinear dependence:

$$\frac{(\eta - 1)}{R_p} - \frac{\eta}{R_c} \equiv 0$$

or

$$\frac{R_c}{R_p} = \frac{\eta}{(\eta - 1)}, \quad (3.21)$$

thus

$$V_{ref} = V_{go}\left(\frac{1}{R_p} - \frac{1}{R_c} \right)(R_{o1} + R_{o2})$$

$$= V_{go}\left(\frac{\eta}{(\eta - 1)} - 1 \right)\frac{(R_{o1} + R_{o2})}{R_c}; \quad (3.22)$$

the TC of the reference voltage has been exactly canceled for any value of η!

In summary, the resistor values are designed to establish the quiescent-current flow through the circuit (R_p), to cancel the logarithmic term of the base-emitter voltage (R_c), to cancel the linear term (R_{o1}), and to set the desired value of the reference voltage (R_{o2}). Figure 3.16(b) is simply a low-voltage version of the one in Figure 3.16 (a). The minimum supply voltage is approximately 2.7 V ($2V_{BE}$ + $V_{GS\text{-Natural}} V_{DS\text{-Natural}}$ + $V_{SG\text{-PMOS}}$ ≈ 1.4 + 0.2 + 0.2 + 0.9 V) and 1.8 V (V_{BE} + $V_{DS\text{-Natural}}$ + $V_{SG\text{-PMOS}}$ ≈ 0.7 + 0.2 + 0.9 V) for the circuits in Figure 3.16(a) and (b), respectively. The cost of reducing the minimum input voltage for version (a) is additional quiescent-current flow; version (b) is less power efficient. The minimum supply voltage varies with temperature, but only slightly, since the TC of V_{BE} (negative) tends to cancel that of V_{SG} (positive). Furthermore, η can be any noninteger value and the circuit can still be designed to yield a theoretically exact reference voltage (potentially to achieve less than 3 ppm/°C).

Table 3.1 shows a qualitative comparison of the different curvature-correcting techniques surveyed. All the techniques can potentially achieve a temperature-drift variation of less than 20 ppm/°C. The table, however, gauges them for their worthiness with respect to quiescent-current flow and headroom limit. These characteristics are especially intrinsic in todays market, where battery-operated circuits play an important role. The driving market, as a result, seeks circuits that have low quiescent-current flow and are operable with low input voltages. The additional number of current-sensitive paths required to transform a first-order into a higher-order topology gauges the category of quiescent-current flow for the purposes of this discussion. The techniques that can potentially yield reference voltages lower than the typical 1.2 V are considered to be low voltage. This categorization

TABLE 3.1. Qualitative Comparison of the Curvature-Correcting Techniques Surveyed

Technique	Quiescent Current	Input Supply Voltage
Temp. dep. resistor ratio	Low	Low
Diode loop	Moderate	Low
β compensation	Low	Moderate
Piecewise linear	Low	Low
Matched: version 1	Moderate	Moderate
Matched: version 2	Moderate	Moderate
Exact: version 1	Low	High
Exact: version 2	Low	Moderate

is used because the fundamental supply voltage limit of a 1.2 V reference is approximately 150 to 300 mV above 1.2 V, which results from typical headroom limitations. Circuits whose inherent headroom limitation restricts operation to input voltages significantly larger than 1.2 V are considered high voltage. As such, the best two topologies for low quiescent-current flow and low voltage are the temperature-dependent resistor ratio technique and the piecewise-linear current-mode method. The second version of the exact method, however, can theoretically achieve the best TC performance with a moderate supply voltage limit of 1.7 V. All these circuits employ current-mode techniques that are relatively simple to realize. Current-mode techniques, in general, reap the benefits of yielding reference voltages lower than 1.2 V.

3.5 SUMMARY

References fall into one of three categories when considering temperature compensation, that is, zero order, first order, or curvature corrected. When a naturally existing voltage is used as a reference without any attempt to cancel its temperature dependence, it is said to have zero-order compensation. When the first-order term, with respect to temperature, is canceled, though, its order increases by 1. Finally, correction of the second-order and higher-order terms constitutes a curvature-corrected reference. Since the electrical characteristics of *p-n* junction diodes are repeatable, predictable, and well characterized over a wide range of currents, diodes are commonly used as the core elements of most first-order and higher-order references; they are the basic naturally existing voltages chosen. Because of its inherent dependence to the bandgap of silicon, they are appropriately called *bandgap references*.

Most high-order precision references compensate the nonlinear temperature-dependent characteristics of the diode by summing an artificially generated nonlinear component. The temperature dependence of this nonlinear term is not necessarily equal to the nonlinearity of the diode voltage, or whatever naturally existing voltage is used to define the reference. The correcting term, for instance, can be proportional to the square of the temperature while the nonlinearity of the diode voltage is actually proportional to a logarithmic expression of temperature ($T \ln T$, where T is temperature). Better performance is attained when the correcting term is designed to mimic the nonlinearity of the diode (i.e., match $T \ln T$). Well, a genre of curvature-correcting circuits goes one step further and attempts to exactly cancel the nonlinearity by deriving the correction component from the diode voltage itself. In the end, though, absolute and exact cancellation is not really possible because of all the parasitic elements present in the circuit. Approximations are consequently made by exploiting the TC of resistors, β (forward-current gain for bipolar transistors), the diode itself, and even the nonidealities of three-terminal devices as when transitioning from the off-region to the on-region.

Due to stringent design constraints in today's system environments, high-order circuits, and even first-order circuits, must also account for the effects of power supply voltage variations as well as of the electrical parasitics present in the integrated circuit and the system at large. Some of these parasitics include resistor tolerance, resistor mismatch, transistor mismatch, leakage current, package drift, power supply noise, and ground bounce. These adverse effects are addressed at every level of design, be it circuit, layout, or trim. The next chapter, as a result, discusses these issues and other practical concerns that must be considered when designing real practical references.

BIBLIOGRAPHY

[1] J.H. Huijsing, et al., *Analog Circuit Design*. The Netherlands: Kluwer Academic Publishers, 1996.

[2] S. Franco, *Design with Operational Amplifiers and Analog Integrated Circuits*. New York: McGraw-Hill, 1988.

[3] *Motorola Linear/Switchmode Voltage Regulator Handbook*, 4th ed., Motorola, 1989.

[4] P.R. Gray and R.G. Meyer, *Analysis and Design of Analog Integrated Circuits*. New York: Wiley, 1993.

[5] R.J. Reay, et. al., "A Micromachined Low-Power Temperature-Regulated Bandgap Voltage Reference," *IEEE Journal of Solid-State Circuits*, vol. 30, no. 12, 1374–1381, December 1995.

[6] P. Gray, "Advanced Analog Integrated Circuits," *Lecture Notes and Diagrams from the University of California at Berkeley*, 1991.

[7] M. Gunawan et al., "A Curvature-Corrected Low-Voltage Bandgap Reference," *IEEE Journal of Solid-State Circuits*, vol. 28, no. 6, pp. 667–670, June 1993.

[8] I. Lee et al., "Exponential Curvature-Compensated BiCMOS Bandgap References," *IEEE Journal of Solid-State Circuits*, vol. 29, no. 11, pp. 1396–1403, November 1994.

[9] G.A. Rincon-Mora and P.E. Allen, "A 1.1 V Current-Mode and Piecewise-Linear Curvature-Corrected Bandgap Reference," *IEEE Journal of Solid-State Circuits*, vol. 33, no. 10, pp. 1551–1554, October 1998.

[10] W.T. Holman, "A New Temperature Compensation Technique for Bandgap Voltage References," *IEEE International Symposium on Circuits and Systems*, vol. 1, pp. 385–388, 1996.

[11] J.H. Huijsing et al., *Low-Noise, Low-Power, Low-Voltage; Mixed-Mode Design with CAD Tools; Voltage, Current and Time References*. The Netherlands: Kluwer Academic Publishers, 1996.

[12] G.M. Meijer et al., "A New Curvature-Corrected Bandgap Reference," *IEEE Journal of Solid-State Circuits*, vol. SC-17, no. 6, pp. 1139–1143, December 1982.

CHAPTER 4

DESIGNING PRECISION
REFERENCE CIRCUITS

A precision reference is not merely a temperature-compensated voltage, or current; it is supposed to be a transiently stable temperature-compensated circuit whose output is impervious to variations in process, supply voltage, load, and noise. Accuracy, as a result, depends on several factors, one of which, of course, is temperature. Well, the dominant trend for the past several years has been toward higher accuracy under an environment that is increasingly more stringent and harsh. Thoroughly integrated mobile battery-operated solutions, like laptops, pagers, cellular phones, and the like, are partially responsible for this market direction. Since mobile units have strict space limitations, the number of batteries and the number of discrete integrated circuits (ICs) are low. The packing density of each IC must therefore increase accordingly to accommodate the space-efficient requirements of the design, which means finer photolithography. Operating voltages are low, as a result, not only because of a reduced number of battery cells but because of lower breakdown voltages, which is the ultimate consequence of a fine layout pitch. Effective dynamic range is also low with lower operating voltages. This effect results because noise floors do not decrease proportionately with supply voltage, a fundamental reality. The effective reduction in dynamic range forces the requirements of the circuit to be stricter [1]. In the end, precision references are in high demand! Though this demand does not, and will not, apply to all applications, its trend has been, and will be, increasing in the future since the thrust of technology is toward higher integration of increasingly complex systems.

The actual design of the reference must cater to the particular needs of the application and the process technology. It must consider process technology, load, systematic noise, input voltage characteristics, and so on. Typically, the load, unlike that of a regulator, will not vary significantly but will still impose significant design constraints on the reference. Systematic noise is another source of concern for it may be directly injected through the load, which transiently affects the accuracy of the reference. Systematic noise may also be injected through the power supply. Sudden changes in the load of the input supply itself, like the ones caused by digital circuitry, cause these transient variations to occur. The power supply rejection (PSR) performance of the circuit determines the vulnerability of the circuit to noise injected through the input supply. Incidentally, steady-state changes in the supply voltage also affect the reference. These effects, in particular, are termed *line regulation effects*, where *line* refers to input supply voltage.

The preceding chapters have mainly dealt with the design of temperature-compensated references. This chapter discusses how to convert these references into precision circuits. Toward this end, a discussion on process-dependent variations and nonidealities is first offered. The induced errors illustrate some of the design challenges encountered in fabricating a real-working integrated reference circuit as well as some of the resulting layout and circuit implications. Ultimately, issues like load and line regulation effects are considered and taken into account when designing the circuit. Examples are presented to aid the reader in understanding the practical issues discussed.

4.1 ERROR SOURCES

As the specification requirements become more stringent, the errors become more critical in a design. These errors are addressed at every level of design (i.e., system, circuit, and layout). Their negative effects on references manifest themselves in the temperature coefficient and in the absolute value of the actual reference voltage (V'_{ref}, voltage value where the temperature-drift performance is at its best). Several factors affect these two characteristics and they include, among others, mirror mismatches, resistors' tolerance, resistors' temperature variation, Early voltage, channel-length modulation, resistor mismatches, transistor mismatches, package drift, and input offset voltages (if operational amplifiers are used).

The extent of the effects of the errors on performance depends on the particular circuit. These inaccuracies are not really as important in zero-order references as they are in higher-order references because the temperature drift is already expansive in low-order references. The parasitic errors become more significant as the order of correction is increased, though. A discussion of all the errors is not possible due to the various reference topologies that exist. However, most of the dominant errors can be traced down to the basic building blocks of the reference (i.e., diode voltage, PTAT current generator, etc.). As such, the following section will discuss the errors associated with a popular topology—the *Brokaw cell.*

4.1.1 Qualitative Effects

The vulnerable areas for temperature-drift performance are the actual diode voltage and the correcting terms that are summed to the diode voltage to generate the reference, i.e., PTAT and $PTAT^2$ (if applicable) voltages. The culprits are the resistors and the transistors—in particular, their mismatch errors, their tolerance over process, and their temperature-drift performance. The area that is most susceptible to these errors is the PTAT current generator and the actual stage that defines the reference voltage. Figure 4.1(a) is a generically common circuit that can be used to study the negative effects of the parasitic effects mentioned. For simplicity, the curvature-correction scheme is ignored (I_{PTAT}^2). A full derivation of the effects of these errors on the actual reference voltage is shown in Appendix A.4. Table 4.1 shows the results of that analysis.

Resistor mismatch refers to relative percent differences in value among resistors that are theoretically equal. This type of error affects the reference by changing the coefficient of the PTAT voltage term, which is the voltage across resistors $R1$ and $R2$. The PTAT voltage gain factor is defined by a resistor ratio, sum ($R1 + R2$) to R. However, this ratio deviates from its ideal value by the mismatch characteristics of the resistors, thereby altering the reference voltage. This percent mismatch can be somewhere between 0.1 and 5%, depending on the material used to create the resistors as well as on their physical layout properties within a given process technology.

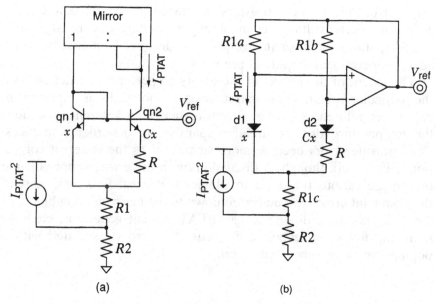

Figure 4.1 Curvature-corrected bandgap circuit.

Tolerance, on the other hand, refers to the absolute percent difference between the theoretical value and the real fabricated value. As such, the tolerance of resistor R also affects the reference. It affects it by changing the absolute value of the PTAT current, which flows through transistor qn1 to define its base-emitter voltage. The base-emitter voltage is a function of the PTAT current, which is a function of the tolerance of resistor R. This tolerance, unfortunarely, can be anywhere between 10 and 30%. It is noteworthy to mention, though, that large changes in current will lead to small changes in base-emitter voltage because of its exponential relationship. As a result, the effects of resistor tolerance are not as dominant among all parasitic.

Mirror mismatches, similar to resistor mismatches, refer to relative percent matching errors between one transistor and others. These kinds of errors ultimately affect the base-emitter voltage of the bandgap circuit core, just like resistor tolerance. The similarity to resistor tolerance is because the current flowing through transistor qn1 is offset from its ideal value by the inaccuracy of the mirror. Errors in the current mirror also affect the PTAT voltage term of the reference, which is, again, the voltage across $R1$ and $R2$. This propagation of the error is because the current flowing through resistors $R1$ and $R2$ is deviated from its ideal value. Its ideal value is the sum of

TABLE 4.1. Parasitic Effects on the Relationship of the Ideal Reference Voltage

Basic Relation:

$$V_{\text{ref}} = V_{BE1} + 2I_{\text{PTAT}}(R1 + R2) \tag{4.1}$$

RELATIVE PERCENT MISMATCH (RESISTOR)

$$(R1 + R2)_x = AR(1 + \delta_{RR}) \qquad V_{\text{ref-}x} = V_{\text{ref}} + 2I_{\text{PTAT}}(R1 + R2)\delta_{RR} \tag{4.2}$$

ABSOLUTE PERCENT MISMATCH (RESISTOR)

$$R_x = R(1 + \delta_{RA}) \qquad V_{\text{ref-}x} \approx V_{\text{ref}} - V_T \delta_{RA} \tag{4.3}$$

MIRROR PERCENT MISMATCH

$$I_{C1} = I_{C2}(1 + \delta_M) \qquad V_{\text{ref-}x} \approx V_{\text{ref}} + V_T \frac{\delta_M}{\ln C}$$

$$+ I_{\text{PTAT}}\left(1 + \frac{\delta_M}{\ln C}\right)\delta_M(R1 + R2) \tag{4.4}$$

NPN MISMATCH

$$x : C(1 + \delta_{\text{NPN}})x \qquad V_{\text{ref-}x} \approx V_{\text{ref}} + V_T \frac{\delta_{\text{NPN}}}{\ln C} + \frac{2V_T \delta_{\text{NPN}}}{R}(R1 + R2) \tag{4.5}$$

NPN EARLY VOLTAGE (FINITE OUTPUT RESISTANCE)

$$I_{\text{PTAT-}x} = \frac{V_T}{R}\ln\left[\frac{C\left(1 + \dfrac{V_{CE2}}{V_A}\right)}{\left(1 + \dfrac{V_{CE1}}{V_A}\right)}\right] \tag{4.6}$$

$$V_{\text{ref-}x} = V_{\text{ref}} + V_T \ln\left(\frac{I_{\text{PTAT-}x}}{I_{\text{PTAT}}}\right)$$

$$+ 2\frac{V_T}{R}(R1 + R2)\ln\left[\frac{\left(1 + \dfrac{V_{CE2}}{V_A}\right)}{\left(1 + \dfrac{V_{CE1}}{V_A}\right)}\right] \tag{4.7}$$

RESISTOR'S TEMPERATURE COEFFICIENT

$$R(T) \approx R(T_r)\left[1 + A(T - T_r) + B(T - T_r)^2\right] \tag{4.8}$$

$$V_{\text{ref-}x} \approx V_{\text{ref}} - V_T \ln\left[1 + A(T - T_r) + B(T - T_r)^2\right] \tag{4.9}$$

two equal PTAT currents; however, the resulting current is actually the sum of two slightly dissimilar currents. The extent of the mismatch depends on how the mirror is built. It can range from 0.5 to 5% mismatch. Bipolar devices generally yield better matching performance than MOS transistors. Both types of devices can improve their matching characteristics, though, by using emitter or source degeneration resistors. Relying on resistors assumes, of course, that the resistors themselves can be laid to yield better matching performance, which is typically a safe assumption. The output resistance of the mirror, its invulnerability to changes in collector or drain voltages, also affects the accuracy of the mirror. (Early voltage and channel-length modulation effects for PNP and PMOS transistors, respectively). A different type of mirror concept can be implemented whereby a resistor load and an operational amplifier is used instead, as shown in Figure 4.1(b). In this latter configuration, the mismatch error would be a function of the resistors and the input offset voltage of the operational amplifier.

The actual matching capabilities of the NPN devices defining the PTAT current—transistor mismatch between qn1 and qn2 or d1 and d2—are intrinsic for precision references. The consequence of mismatch in these two devices is an offset voltage. In other words, given the same current density and collector-emitter voltage, the base-emitter voltage of these devices differs by an effective offset voltage. This voltage is superimposed across resistor R. As a result, the PTAT current changes, thereby altering the base-emitter voltage as well as the voltage across resistors $R1$ and $R2$. The offset voltage can range from 0.5 to 5 mV depending on the physical layout and process constraints. The Early voltage of these devices similarly affects the reference. Given equal current densities and different collector-emitter voltages, the base-emitter voltage of two devices will differ. As a result, an error in the PTAT current is generated, which similarly affects the base-emitter voltage as well as the voltage across $R1$ and $R2$ (PTAT voltage).

Assuming that all resistors are fabricated with the same material, the temperature coefficient (TC) of resistors $R1$ and $R2$ cancels the TC of R. They essentially cancel because sum ($R1 + R2$) is divided by R, the denominator within the PTAT current relationship. However, the temperature coefficient of resistor R does affect the base-emitter voltage. In fact, the current flowing through the collector is not truly PTAT because of this vulnerability to R. In other words, the TC of resistor R affects the TC of the PTAT current. Since the base-emitter voltage is a function of the PTAT current, the TC of the base-emitter voltage is also affected.

Example 4.1: What is the total effect of the following on the reference voltage of Figure 4.1(a)?

1. $\pm 1\%$ resistor mismatch (δ_{RR}),
2. $\pm 20\%$ resistor tolerance (δ_{RA}),
3. $\pm 5\%$ current mirror mismatch (δ_M),
4. $\pm 2\%$ NPN transistor mismatch (δ_{NPN}),
5. 50 V NPN transistor Early voltage (V_A), and
6. 500 ppm/°C resistor TC with a 0 ppm/°C^2 quadratic dependence.

Assume that $C = 10$, $V_T = 0.0258$ V, $V_{CE2} = 4$ V, $V_{CE1} = V_{BE1} \approx 0.65$ V, $(R1 + R2) \div R = 5$, and only first-order correction is implemented ($I_{PTAT^2} = 0$).

1. Resistor mismatch:

$$V_{\text{error-1}} = 2\left(\frac{V_T \ln C}{R}\right)(R1 + R2)\delta_{RR} \approx \pm 5.9 \frac{T}{300 \text{ °K}} \text{ mV}$$

2. Resistor tolerance:

$$V_{\text{error-2}} = -V_T \delta_{RA} \approx \pm 5.2 \frac{T}{300 \text{ °K}} \text{ mV}$$

3. Current mirror mismatch:

$$V_{\text{error-3}} \approx V_T \frac{\delta_M}{\ln C} + \left(\frac{V_T \ln C}{R}\right)\left(1 + \frac{\delta_M}{\ln C}\right)\delta_M(R1 + R2)$$

$$\approx \begin{cases} +15.7 \dfrac{T}{300 \text{ °K}} \text{ mV} \\ -15.1 \dfrac{T}{300 \text{ °K}} \text{ mV} \end{cases}$$

4. NPN transistor offset voltage:

$$V_{\text{error-4}} \approx V_T \frac{\delta_{\text{NPN}}}{\ln C} + \frac{2V_T \delta_{\text{NPN}}}{R}(R1 + R2) \approx \pm 5.4 \frac{T}{300\,°\text{K}}\ \text{mV}$$

5. Early voltage:

$$I_{\text{PTAT-err}} = \frac{V_T}{R} \ln \left[\frac{C\left(1 + \dfrac{V_{CE2}}{V_A}\right)}{\left(1 + \dfrac{V_{CE1}}{V_A}\right)} \right] \approx \frac{61T}{R(300\,°\text{K})}\ \text{mA}$$

Thus,

$$V_{\text{error-5}} = V_T \ln \frac{I_{\text{PTAT-err}}}{\left(\dfrac{V_T \ln C}{R}\right)} + 2\frac{V_T}{R}(R1 + R2) \ln \frac{\left(1 + \dfrac{V_{CE2}}{V_A}\right)}{\left(1 + \dfrac{V_{CE1}}{V_A}\right)}$$

$$\approx 17.2 \frac{T}{300\,°\text{K}}\ \text{mV}$$

6. Resistors' TC:

$$V_{\text{error-6}} = -V_T \ln\left[1 + A(T - T_r) + B(T - T_r)^2\right]$$

$$\approx -25.8 \frac{T}{300\,°\text{K}} \ln[1 + 0.0005(T - 300\,°\text{K})]\ \text{mV}$$

Thus,

$$V_{\text{error}}^*(T_r) = \sqrt{\sum_{n=1}^{6} V_{\text{error-}n}(T_r)^2}$$

$$\approx \sqrt{5.9^2 + 5.2^2 + 15.7^2 + 5.4^2 + 17.2^2}\ \text{mV} \approx 25.2\ \text{mV}$$

It is evident that Early voltage of the NPN transistors ($V_{\text{error-5}}$) and mismatch error of the current mirror ($V_{\text{error-3}}$) have the biggest impact on the reference voltage. Some errors like resistors' TC, Early voltage, and lambda (channel-length modulation parameter) effects can be readily modeled by most simulators. Applying equivalent offset voltages can also simulate tolerance and mismatch errors. For instance, a transistor mismatch error can be translated into an equivalent voltage offset error, which, in turn, is superimposed and added to the circuit. This offset voltage macromodel should exhibit the appropriate temperature dependence, derived in Appendix A.4 with equation (A.4.15). Ultimately, painstaking effort is dedicated to the design of the mirror and the overall physical layout of the circuit. It is important to note that most of the errors have a dependence to temperature and all tend to be either PTAT or CTAT in nature, assuming that the errors themselves are not temperature-dependent. Errors such as resistor tolerance δ_{RA}, resistor mismatch δ_{RR}, mirror mismatch δ_M, and NPN transistor mismatch δ_{NPN} are mostly a function of process and associated constraints in photolithography and therefore tend to have little dependence on temperature. As a result of this tendency, the variant factor in a trimming algorithm should be the PTAT component! The trimming resistor for the circuit at hand should therefore be $R2$ (assuming that I_{PTAT^2} is zero). Early voltage and lambda effects may exhibit some variation over temperature but this may be included in the model of the appropriate transistor. In the end, however, a full characterization of the circuit may be necessary to reveal the best operating point of the reference voltage, value for which V_{ref} exhibits the best temperature drift performance.

4.2 THE OUTPUT STAGE

The output stage of the reference is designed to cater to the demands of the load for a given application. As such, the circuit may take one of several forms ranging from a purely voltage-mode or current-mode to a mixed-mode topology. Within the present context, *mixed mode* refers to circuits employing both voltage-mode and current-mode circuit techniques. Additionally, the output of the reference can either be regulated or unregulated, depending, again, on the loading demands of the system. In the end, the load and, of course, headroom limits define the output structure that is best suited for a given design. For instance, an unregulated current-mode output stage may be acceptable for a design that does not require a low-impedance output.

If the load of the output is a large capacitor and its demand for steady-state current is nonexistent, there may not be a need for a low-impedance output. In this case, noise generated by other circuits is effectively shunted to ground via the large capacitor, thereby not degrading the accuracy performance of the reference during transient load-current events. Finally, the output stage is also dependent on the nature of the reference circuit itself. A system may not demand a regulated output, but many circuit topologies inherently require a regulating loop to establish the reference voltage. The following sections address the design considerations and limitations of each type of output stage.

4.2.1 Voltage-Mode

Historically, a voltage-mode output has been the most common technique used in reference circuits. This tendency results because the basic temperature-dependent element within the core of reference circuits is typically a voltage (i.e., Zener diode voltage). Currents are, for the most part, derived from these characteristic voltages. A voltage-mode output, as it pertains to references, is characterized by the sum of temperature-dependent *voltages*. Most zero-order, first-order, and even higher-order bandgap or Zener references adopt a voltage-mode output stage. Figure 4.2(a) illustrates the typical output structures of first-order Zener and bandgap references. The positive temperature coefficient (TC) of the Zener diode or, alternatively, the positive TC of the proportional-to-absolute temperature (PTAT) voltage roughly cancels the negative TC of the forward-biased diode voltage.

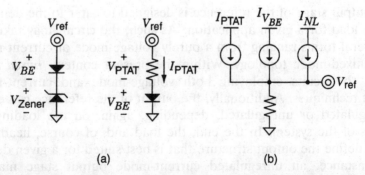

(a)   (b)

Figure 4.2 Output structures: (a) voltage-mode and (b) current-mode.

The voltage-mode approach has the advantage of being simple. In the case of bandgap references, the output is trimmed by merely changing the size of a resistor. This introduces a PTAT trim voltage, which is the general dependence of the offsets in the circuit as discussed earlier in Section 4.1. Moreover, the basic temperature component of the circuit (the diode voltage) is used directly in series with the output voltage, thereby not introducing additional errors. In the end, the diode voltage itself, not its derivative, is used in the output to generate the reference voltage.

The market is driving power supply voltages down, thereby creating lower headroom limits. The thrust for this trend is driven by low-voltage applications as exemplified by battery-operated products. As a result, typical Zener references are not suitable for this environment because of their inherent high operating voltages (i.e., between 5 and 7 V). Similarly, bandgap references are appropriate for input supply voltages exceeding approximately 1.5 V, which corresponds to 300 mV (MOS transistors drain-source or bipolar devices collector-emitter voltage) above a bandgap voltage of 1.2 V. This latter limitation does not present a problem for most systems whose present market demand is driven by battery and breakdown voltages greater than 2.5 V. However, single battery-cell operation and lower breakdown voltages will eventually require the reference to operate at supply voltages of roughly 1 V. A typical single nickel-cadmium (NiCd) or nickel-metal-hydride (NiMH) battery cell has 1.5 V at its output but its voltage will decay to about 0.9 V before totally collapsing [2]. In applications that use such a battery supply, a reference voltage between 0.5 and 0.7 V is desired. Unfortunately, the fundamental limit of the bandgap voltage-mode topology is 1.2 V.

4.2.2 Current-Mode

Reference voltages lower than 1.2 V can be achieved by implementing a current-mode approach. The technique, unlike its voltage-mode predecessor, relies on summing temperature-dependent currents into a resistor as shown in Figure 4.2(b). The current components, as is the case for most curvature-corrected bandgap references, are PTAT, base-emitter voltage derived, and nonlinear (I_{NL}). The product of the sum of the currents and the resistor determines the value of the output voltage. Consequently, the output voltage for this configuration is sufficiently flexible to accommodate a wide range of values ranging from a few millivolts to several volts. For this case, the PTAT current is typically trimmed instead of the output resistor to ensure a PTAT trim voltage. An example of the current-mode output stage is realized and embodied in the bipolar low-voltage reference design shown in Figure 3.9 of Chapter 3. When designed properly, the temperature dependence of the resistor is mostly canceled by the inherent nature of the currents. The currents are inversely proportional to resistors whose TC is equal to the output resistors, assuming they have been fabricated with the same material. Thus, the resulting reference voltage is dependent on a resistor ratio whose TC is independent of the resistors' TC. Appendix B.4 illustrates a mathematical derivation that supports the aforementioned claim. However, low-TC resistors are still preferred because of the adverse effects on the basic generation of the PTAT current and the base-emitter voltage, as discussed in Section 4.1.

4.2.3 Mixed-Mode

The benefits of a current-mode and a voltage-mode topology can be combined to generate a widely flexible structure. Such a structure is illustrated in Figure 4.3. The resulting architecture complements the basic current-mode topology with a voltage-mode resistor ladder. In essence, currents and voltages are summed to generate the output reference voltage. The current-mode approach offers the possibility of lowering the value of the reference voltage while the voltage-mode ladder provides enhanced flexibility for temperature compensation. A low-voltage reference is therefore realized whereby the individual temperature components can be further optimized during the trimming process. The relationship of the reference voltage for this case is

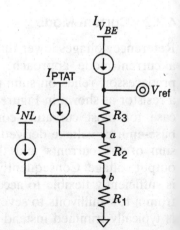

Figure 4.3 Mixed-mode (both voltage-mode and current-mode) output topology for references.

described by

$$V_{\text{ref}} = I_{V_{BE}}(R_1 + R_2 + R_3) + I_{\text{PTAT}}(R_1 + R_2) + I_{NL}R_1, \quad (4.10)$$

where $I_{V_{BE}}$, I_{PTAT}, and I_{NL} correspond to the base-emitter, the PTAT, and the nonlinear temperature-dependent currents respectively.

4.2.4 Regulated versus Unregulated References

A regulated reference uses a feedback loop to minimize the variations of the output voltage with respect to its load and its operating conditions. An unregulated reference, on the other hand, is more vulnerable to variations in the load, such as the case for changes in load current. Essentially, the output resistance of an unregulated circuit is relatively high. If large variations in load current still cause small changes in output voltage, the circuit is said to be a *regulator* and not a *reference*. In summary, a circuit that generates a stable voltage and (1) can only supply currents of less than 1 μA is a reference, (2) can supply currents between 1 and 100 μA is a regulated reference, and (3) can supply currents greater than 100 μA is a regulator. This is how a regulator circuit is differentiated from that of a reference. The regulated reference appropriately belongs to the family of references because its regulating loop is integrated within the bandgap cell itself.

Most integrated systems adopt a regulated reference topology to minimize the effects of noise injection. The low impedance associated with the regulated output makes the reference voltage relatively insensitive to broad-spectrum noise as well as insensitive to systematic noise transients. The fact that it can provide steady-state current is not typically the overriding concern in these applications. Though they can provide some current, they are still not used as regulators because the load-current range is larger and the load-impedance range is lower for applications where a regulator is warranted. Regulators are typically required to be stable over a broad load-capacitor range and over a large equivalent series-resistance (ESR) range. The ESR is the series-parasitic Ohmic resistance that capacitors exhibit. Moreover, the value of the capacitor and the ESR both change with temperature. Unlike regulator loads, though, most regulated references have an invariant output-load environment.

As alluded in the previous discussion, the unregulated reference will exhibit a significant variation in its output voltage for small changes in current. This quantifiable variation not only limits the output current capabilities of the circuit but it also partially defines its vulnerability to noise and, in general, to transient events. The output can only drive circuits with considerably higher input resistance, like the gates of MOS transistors. Moreover, the size of the load-capacitance determines the extent to which the output voltage varies when a transient event occurs. The reference voltage changes less, of course, with increasing load-capacitance. A regulated reference, on the other hand, imposes less stringent requirements on the system because it can drive larger output currents (gates of MOS devices and bases of bipolar transistors) and it exhibits lower output resistance. A lower impedance requirement relaxes the load-capacitance restrictions of the system.

A regulated reference is an unregulated reference with a negative feedback loop. The open-loop gain of this feedback path is relatively high. The resulting closed-loop output resistance $R_{o\text{-}CL}$ is inversely proportional to the open-loop gain A_{OL},

$$R_{o\text{-}CL} = \frac{R_{o\text{-}CL}}{1 + A_{OL}B}, \qquad (4.11)$$

where $R_{o\text{-}OL}$ is the open-loop output resistance and B is the feedback-gain factor. The feedback loop is realized either by integrating it to the basic reference cell or by buffering the output with an operational amplifier in unity-gain configuration. Figure 4.4 illustrates an example of an integrated loop topology and the general circuit configuration for a buffered reference.

The specific integrated loop shown in Figure 4.4(a) actually has two feedback paths, one positive and one negative. The proper configuration ensures that the negative feedback has an open-loop gain greater than that of the positive feedback path. The positive feedback path goes through the collector of an NPN transistor (qn1) that is emitter-degenerated with a resistor. The resistor reduces the effective transconductance of transistor qn1, thereby lowering the resulting open-loop gain. As a result, the polarity of the amplifier must be configured such that the negative feedback path goes through transistor qn2, whose transconductance is not degraded. The loop, of course, must be designed to be stable and may require one or two compensating capacitors. The buffered version of the regulated reference shown in Figure 4.4(b) cascades an operational amplifier in unity-gain configuration to the unregulated reference. This latter circuit is relatively

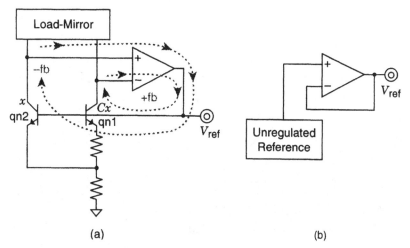

(a)   (b)

Figure 4.4   Regulated reference topologies.

easier to design since the buffer and its compensation requirements are independent of the reference cell. This approach may demand more silicon area and more quiescent-current flow than the integrated loop version. Integrated loops take advantage of already-available devices to formulate and bias the feedback path.

Design Example 4.2: Design a regulated, voltage-mode, first-order bandgap reference for a temperature ranging from 0 to 125 °C, assuming that the bandgap voltage (V_{go}) is 1.2 V and the process constant η is approximately 3.6. Also assume that a biCMOS process is to be used where natural MOS devices are available. The input power supply of the circuit is to be greater than or equal to 2.5 V.

Since the supply voltage is 2.5 V and a voltage-mode topology is desired, the circuit topology illustrated in Figure 4.5 is adopted. Devices qn11, qn12, RA, and RB comprise the basic first-order bandgap reference where the diode voltage across the base-emitter junction of qn11 is summed with the PTAT voltage across resistor RA. Devices $R21$, $R22$, mp21, mp22, mn21, mn22, mn23, mn24, Cc, RO, and qno make up the feedback-error amplifier performing the function of the amplifier in Figure 4.4. Resistor $R12$ and capacitor Cf are used to filter the positive feedback loop with the intent to reduce its gain relative to the negative feedback loop. The purpose of resistor $R11$ is to ensure that the bases of qn11 and qn12 exhibit the same voltage; voltages across $R11$ and $R12$ are equal assuming that the base currents match. Finally, mpi, qnb, rnb, mnb, mpb, and Rpb comprise

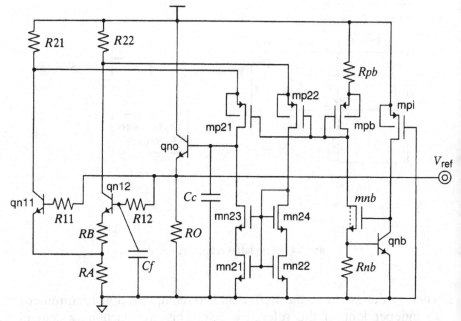

Figure 4.5 Regulated, voltage-mode, first-order biCMOS bandgap reference.

the biasing circuit. Transistor mpi is used as an inaccurate current source. Transistors qnb and mnb along with resistor Rnb derive a predictable CTAT current that is ultimately used to bias the gates of mp21 and mp22 via mpb and Rpb. The circuit, by the way, is self-starting since mpi ensures that the gates of mp21 and mp22 are biased, thereby preventing a zero-current state from occurring.

The first portion of the circuit to be designed is the basic bandgap cell. The same methodology used to design the circuit shown in Example 3.2 will be used to design this circuit. Transistor qn11 is therefore chosen to have an area equal to four times larger than the minimum possible size for an NPN device. A nonminimum device size is chosen to maximize the matching capabilities between transistors qn11 and qn12. Now, the current through qn11 is arbitrarily chosen to be 15 μA. A lower current can be chosen, of course, but it typically makes the circuit more susceptible to substrate and input supply noise. The size of qn11, being necessarily larger than qn12, is chosen to be 8 times larger than qn2. At this point, resistor RB is simply

$$RB = \frac{V_T \ln C}{I_{PTAT}} = \frac{V_T \ln C}{I_{qn11}} = \frac{(25.8m)\ln 8}{15 \ \mu A} \approx 3,580 \ \Omega.$$

The reference voltage is equal to the sum of base-emitter voltage V_{BE11} and the PTAT voltage across resistor RA. The way to design the value of this resistor is to make a first-order approximation of the base-emitter voltage and to determine the PTAT voltage required to produce a zero-TC response. The approximation will neglect the higher-order nonlinear components of the base-emitter voltage,

$$V_{BE11} \approx \left[V_{go} + (\eta - 1)V_{T_r} \right] - \left[V_{go} - V_{BE11}(T_r) + (\eta - 1)V_{T_r} \right] \frac{T}{T_r},$$

where V_{T_r} is the thermal voltage at room temperature T_r and 1 is substituted for the constant x in equation (1.7) of Chapter 1 since the collector current is PTAT ($I_{C2} \propto T^1$). For the voltage reference to have zero TC, its first derivative must be equal to zero, i.e.,

$$\frac{\partial V_{ref}}{\partial T} \bigg|_{T=T_r} = \frac{\partial (V_{BE11} + V_{PTAT})}{\partial T} \bigg|_{T=T_r}$$

$$= -\left[\frac{V_{go} - V_{BE11}(T_r) + (\eta - 1)V_{T_r}}{T_r} \right]$$

$$+ \left[2\frac{I_{PTAT}(T_r)}{T_r} \right] RA \equiv 0,$$

where V_{PTAT} is the PTAT voltage across resistor RA. The multiplier "2" in the above relation comes from the fact that the current through RA is equal to $2I_{PTAT}$. Consequently, the value for RA can be found to be 19.9 kΩ,

$$RA = \frac{V_{go} - V_{BE11}(T_r) + (\eta - 1)V_{T_r}}{2I_{PTAT}(T_r)}$$

$$= \frac{1.2 \text{ V} - 0.67 \text{ V} + 67.1 \text{ mV}}{2(15 \text{ } \mu\text{A})} \approx 19,903 \text{ } \Omega,$$

where V_{BE11} is assumed to be 0.67 V at room temperature. The resulting reference voltage at room temperature is

$$V_{ref}(T_r) = V_{BE}(T_r) + V_{PTAT}(T_r)$$

$$= 0.67 \text{ V} + 2(15 \ \mu\text{A})(19.9 \text{ k}\Omega) \approx 1.267 \text{ V}.$$

The feedback amplifier is now designed. The voltage across resistors $R21$ and $R22$ is chosen to be 200 mV. Since the input supply voltage is greater than 2.5 V, a low voltage drop is required to ensure that mp21 stays in saturation. Choosing, for ease of design and since no limitations were given to quiescent current, the current through mp21 and mp22 to also be 15 μA forces resistors $R11$ and $R12$ to be 6.7 kΩ each:

$$R11 = R12 = \frac{200 \text{ mV}}{I_{PTAT} + I_{mp21}} = \frac{200 \text{ mV}}{15 \ \mu\text{A} + 15 \ \mu\text{A}} = 6,667 \ \Omega.$$

The voltage at the drain of mp21 could reach up to roughly 2 V at low temperatures ($V_{ref} + V_{BE}$). Devices mp21 and mp21, as a result, are designed to have a saturation voltage of less than 300 mV, which yields a device size of 120/5 where a channel length of five microns, assuming a minimum size of one micron, is chosen to minimize channel-length modulation effects,

$$V_{mp21\text{-sat}} = \sqrt{\frac{2I_{mp21}}{K'(W/L)_{mp4}}} = \sqrt{\frac{2(15 \ \mu\text{A})}{15\mu(W/L)_{mp4}}} \leq 0.3 \text{ V}$$

or

$$(W/L)_{mp21} \geq 22.2,$$

where transconductance parameter K' for PMOS devices is assumed to be 15 μA/V^2.

Devices mn21 through mn24 comprise the load mirror of the amplifier. Devices mn23 and mn24, in particular, are natural NMOS devices. As such, they have a threshold voltage of zero volts, nominally. This low-threshold voltage is exploited by sharing the gate connection with mn21 and mn22. In the end, sufficient headroom is left across mn21 and mn22 to operate in saturation and no additional devices were necessary to bias the gates of mn23 and mn24. Since the voltage at the drain of mn23 is going to be greater than approximately 1.7 V ($V_{ref} + V_{BE}$), the saturation voltage of mn23/mn24 and mn21/mn22 is chosen to be 500 mV, which leaves ample headroom,

$$V_{mn21\text{-}sat} = \sqrt{\frac{2I_{mp21}}{K'(W/L)_{mn21}}} = \sqrt{\frac{2(15\ \mu A)}{50\mu(W/L)_{mn21}}} \leq 0.5\ V$$

or

$$(W/L)_{mn21} \geq 2.4,$$

where transconductance parameter K' for all NMOS devices is assumed to be 50 $\mu A/V^2$. Devices mn21 and mn22 need to match well to produce low offset voltages. As a result, their channel length is chosen to be 25 m. On the other hand, mn23 and mn24 are not as critical and can therefore tolerate a shorter channel length (i.e., 10 μm). Consequently, mn21, mn22, mn23, and mn24 have a device size of 65/25, 65/25, 25/10, and 25/10, respectively.

Finally, qno is chosen to be minimum size since it does not demand any accuracy requirements and its current load is low. Its quiescent current, though, is arbitrarily chosen to be 5 μA. Collector current ensures a more predictable response, not to mention a faster reaction. Resistor *RO* is therefore designed to be 255 kΩ,

$$RO = \frac{V_{ref}}{5\ \mu A} = \frac{1.267\ V}{5\ \mu A} = 253.4\ k\Omega.$$

Capacitor *Cc* is added to compensate the amplifier. The circuit itself may not need this compensating capacitor since the dominant pole of the amplifier is inherently the base of qno. All other nodes in the feedback path have low impedance (i.e., source of mp21/mp22 and emitter of qno). As such, a relatively small capacitor value is chosen (i.e., 10 pF). A larger capacitor may be warranted if enough capacitance is added to the output of the circuit. Resistor/capacitor combination *R12/Cf* is chosen to yield a positive feedback pole frequency of 100 kHz, which roughly translates to 160 kΩ and 10 pF,

$$f_{+\text{feedback-pole}} = \frac{1}{2\pi R12Cf} = \frac{1}{2\pi(160)(10p)} = 99.5\ k\Omega.$$

Resistor $R11$, of course, is designed to match $R12$, thereby also having a value of 160 kΩ. It is noted, at this point, that the reference voltage is also a function of the base current of qn11 or, equivalently, a function of β. The simulator is used to ultimately center the PTAT-to-diode voltage ratio while taking into account finite β errors.

The last section of the circuit to be designed is the biasing circuit. Device mpi is simply used as a current source and it is arbitrarily chosen to be roughly 1 μA at an input supply voltage of 2.5 V, $W/L = 2/50$,

$$I_{mpi} = \frac{K'W}{2L}\left[2\left(V_{SG\text{-}mpi} - |V_{tp}|\right)V_{SD\text{-}mpi} - V_{SD\text{-}mpi}^2\right]$$

$$= \frac{(15\mu)W}{2L}\left[2(2.5 - 0.7)(1.23) - 1.23^2\right] \equiv 1\ \mu A$$

or

$$(W/L)_{mpi} = \frac{1}{22},$$

where the threshold voltage $|V_{tp}|$ is assumed to be 0.7 V and the source-drain voltage is assumed to be 1.23 V (2.5 V $- V_{mnb\text{-sat}} - V_{BE\text{-qnb}}$, where $V_{mnb\text{-sat}}$ is less than 0.6 V). It is noted that mpi is assumed to be nonsaturated since its source-drain voltage is nominally less than the difference between its source-gate voltage (equal to the supply voltage) and its threshold voltage. Now, qnb is chosen to be minimum size and Rnb is designed to conduct, nominally, 5 μA; thus, Rnb is designed to be 134 kΩ, $I_{Rnb} = V_{BE}/Rnb$. Mnb is designed to have a minimum channel length (3 μm) and a width of 5 μm, which yields a saturation voltage of roughly 350 mV. Mpb, having a current density of roughly one-third the current flowing through mp21/mp22, is designed to be one-third in size and matched (i.e., 40/5). The voltage across Rpb should match the voltage across $R21/R22$; thus, its size is chosen to be 40 kΩ, 200 mV $= (5\ \mu A) \times Rpb$.

At this point, the circuit is simulated to determine its temperature-drift performance, closed-loop transient response, and steady-state performance. Startup should also be tested, in a steady-state test, by slowly ramping up the supply and, in a transient test, by rapidly pulsing the supply once. In the transient run, the reference should recover to its steady-state value.

4.3 DESIGNING FOR POWER SUPPLY REJECTION AND LINE REGULATION

The overall objective of designing a precision reference is to achieve high accuracy over all working conditions. The load, the input voltage, and the system itself define the operating limits of the circuit. The output stage is designed according to the loading requirements of the system but the overall circuit must also be designed to accommodate the demands of the input voltage. A voltage variation extending from 5 to 10 V on the input power supply voltage is not uncommon for many integrated circuits. The resulting deviation in the reference voltage due to this supply variation is described by the term *line regulation*. More explicitly, line regulation performance refers to the steady-state voltage changes in the reference resulting from DC changes in the input supply voltage.

The supply voltage can also be a source of transient noise. The environment where the circuit lies dictates the severity of the noise injected through the input supply. This noise may be random or systematic in nature, having frequency components encompassing a broad spectrum. The robustness and the stability of the circuit with respect to this noise are gauged by the power supply rejection (PSR) parameter. The term *PSR*, for the case of references, applies to the changes in the reference voltage that result from variations in the input supply. In practical circuits, the frequency response PSR performance is not the same throughout the input voltage range. Early voltage and channel-length modulation effects typically cause this PSR variation. In other words, the transistors' output impedance changes with steady-state biasing conditions (drain-source or collector-emitter voltages). Though PSR refers to small signal response, the general techniques used to improve line regulation will also apply to PSR. In summary, the overall design of the reference must address the issue of a variable power supply voltage. Toward this end, circuits typically employ cascoding and/or preregulating techniques to diminish the effects of a changing supply.

4.3.1 Cascodes

The basic concept behind improving power supply rejection (PSR) and line regulation performance pivots around increasing the effective impedance from sensitive nodes (especially the reference voltage) to the input power supply voltage. The concept of PSR may be described

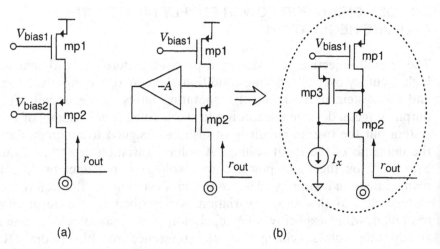

(a)   (b)

Figure 4.6 Cascoding techniques used to increase the impedance to the power supply.

by assuming an impedance divider from the power supply to ground and whose common-middle node is the reference voltage. The effective PSR is expressed as

$$\text{PSR} = \frac{\Delta V_{in}}{\Delta V_{ref}} = \frac{z_{gnd} + z_{in}}{z_{gnd}}, \tag{4.12}$$

where ΔV_{ref} and ΔV_{in} refer to changes in the reference voltage and the input supply voltage while z_{gnd} and z_{in} represent the effective impedance from the reference to the ground node and to the input supply voltage, respectively. Consequently, PSR performance improves (becomes larger) as the impedance from the reference voltage to the input supply increases. Cascodes can therefore be used to increase the output impedance of a particular node to the input supply. Incidentally, PSR is also sometimes expressed as the reciprocal of equation (4.12). This text, however, will consistently refer to PSR as described in the aforementioned equation.

Figure 4.6 illustrates how cascoding techniques can be applied to a particular device (transistor mp1). If the cascode device (transistor mp2) were not present, only the channel-length modulation effects of transistor mp1 (R_{SD1}) define the output resistance. The simple ("unregulated") cascode circuit of Figure 4.6(a) has a significantly larger output resistance. Its output resistance (R_{out}) is, as derived from equation (1.16) in the current mirror section of Chapter 1 where gain A is equal to 1,

$$R_{\text{out}} \approx (1 + g_{m2}R_{SD1})R_{SD2} + R_{SD1} \approx g_{m2}R_{SD1}R_{SD2}, \quad (4.13)$$

where R_{SD1} and R_{SD2} are the output resistances of devices mp1 and mp2 while g_{m2} is the transconductance of transistor mp2. The resulting output resistance, relative to a standalone transistor, can be approximately between 30 and 50 dB greater. The regulated cascode circuit shown in Figure 4.6 (b) exhibits even larger output resistance. A loop with negative feedback gain is added around the cascoding device to *regulate* the output current, thereby further increasing the effective output resistance of transistor mp1, as derived in Chapter 1. The resulting relation is

$$R_{\text{out}} \approx A g_{m2}R_{SD1}R_{SD2} \approx g_{m3}R_{SD3}g_{m2}R_{SD1}R_{SD2}, \quad (4.14)$$

where A is the voltage gain $(g_{m3}R_{SD3})$, g_{m3} is the transconductance of transistor mp3, and R_{SD3} corresponds to the output resistance of the same device. It is assumed, for the above relationship, that the output resistance of the current source I_x is an order of magnitude greater than R_{SD3}. However, if this impedance were to be lower, R_{SD3} in equation (4.14) would simply change to the effective parallel impedance of R_{SD3} and the output resistance of current source I_x. The overall output resistance of the circuit is at least an order of magnitude greater than the unregulated cascode case.

4.3.2 Pseudo-Power Supply

Another method of improving power supply rejection (PSR) is by preregulating the input voltage. The noisy and variable supply is essentially isolated from the reference by means of a buffer. As a result, the variations of the preregulated voltage, pseudo-supply for the reference circuit, are much lower. For the purposes of analyses, the preregulator is described by an effective output impedance from the pseudo-supply to the input voltage $(z_{\text{pseudo-in}})$ and by another impedance from the pseudo-supply to ground $(z_{\text{pseudo-gnd}})$. Consequently, the PSR of the overall circuit, as illustrated in Figure 4.7 (a),

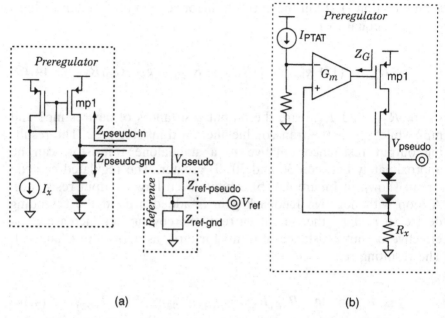

(a)   (b)

Figure 4.7 Preregulated pseudo-supply voltages for reference circuits.

is

$$\text{PSR} = \text{PSR}_{\text{pre-reg}} \cdot \text{PSR}_{\text{regerence}}$$

$$= \frac{z_{\text{pseudo-in}} + z_{\text{pseudo-gnd}}}{z_{\text{pseudo-gnd}}} \cdot \frac{z_{\text{ref-pseudo}} + z_{\text{ref-gnd}}}{z_{\text{ref-gnd}}}, \quad (4.15)$$

where $\text{PSR}_{\text{pre-reg}}$ and $\text{PSR}_{\text{reference}}$ refer to the PSR performance of the preregulator and the reference circuit itself while $z_{\text{ref-gnd}}$ and $z_{\text{ref-pseudo}}$ are the impedance values from the reference to ground and from the reference to the pseudo–power supply, respectively. For best results, the impedance from the pseudo-supply to the input supply must be large and the impedance from the pseudo-supply to ground must be low. The simple circuit shown in Figure 4.7(a), enclosed within the preregulator boundary, achieves this condition. The three diodes define the low impedance to ground ($3r_{be}$) while the drain-source impedance of mp1 determines the impedance to the input supply (r_{sd1}). For further improvements in PSR, the cascoding techniques illustrated in Figure 4.6 can be implemented to increase the impedance from the pseudo-supply to the input supply. Additionally, the pseudo-supply can be regulated with a negative feedback loop to decrease the impedance from the pseudo-supply to ground as shown in Figure 4.7(b), which is a variation of the circuit designed in [3] and [4],

$$
z_{\text{pseudo-gnd}} = \frac{2r_{b_e} + R_x}{1 + G_m Z_G g_{mp1} R_x},
\qquad (4.16)
$$

where Z_G is the impedance at the output of transconductor G_m and g_{mp1} is the transconductance of transistor mp1. The Ohmic voltage summed with the two diodes is PTAT to partially compensate the negative temperature coefficient of the diode voltage, thus potentially obtaining a first-order temperature-compensated pseudo-supply. It is not intrinsic for the preregulated supply to be a first-order reference, though; it could simply be a zero-order reference. The PSR of the reference itself is sufficiently large to dampen the effects of the temperature drift of the pseudo-supply. Consequently, alternate realizations of the feedback loop can be formulated. In the end, the tradeoff of enhancing PSR performance is additional quiescent-current flow and silicon area. The total variation tolerated by the system determines if the added level of complexity is necessary to mitigate the effects of a variable input supply.

The minimum steady-state input voltage allowed for proper operation (headroom limit) is increased if a preregulator is added, unfortunately. This additional overhead is typically between 100 and 400 mV. The PSR and the line regulation performance of the device are best when the input supply is above the pseudo-supply voltage by, at least, roughly 400 mV. This characteristic results because the series device between the input and the pseudo-supply displays its largest output resistance when operated in the saturated region (MOS transistors) or forward-active region (bipolar devices). The characteristic impedance of the series element decreases significantly as it enters the linear region, which happens when the voltage between the input and the pseudo-supply is low. This differential voltage is called the *dropout voltage*. During dropout, line regulation performance and PSR degrade. Figure 4.8 shows the experimental line regulation results of a bandgap reference with and without a preregulated pseudo-supply voltage. The reference variation over input voltage range decreases from 175 to 1.8 mV/9V with a preregulated pseudo-supply [5].

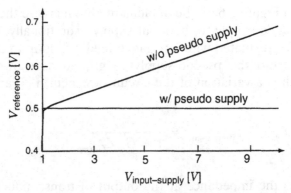

Figure 4.8   Line regulation performance of a bandgap reference with and without a preregulated pseudo-supply.

Design Example 4.3: Design a preregulator whose input voltage ranges from 1.5 to 10 V and whose load is a reference circuit having between 30 and 40 μA of quiescent-current flow. The reference circuit can tolerate a minimum pseudo-supply voltage of 1.3 V.

Figure 4.9 illustrates a circuit that meets the aforementioned requirements. The voltage of the pseudo-supply is defined with three series diode-connected devices. Two diode voltages would have yielded less than 1.3 V at high temperatures. The negative feedback loop of the preregulator is comprised of transistors qn1, mp1 through mp3, and the three diode-connected transistors, qn3 through qn5. Transistor qn2 is used as a current source for mp2. It also provides a positive feedback path to the output; however, its gain is lower than the corresponding negative feedback gain. The relatively low positive feedback gain results because transistor qn2 is emitter degenerated with a resistor while qn1 is not. Transistors mps1 through mps3 are used to "startup" the circuit. Transistor mps2 is used to sense the current through qn1 to ultimately compare it against current source mps1 (diode-connected device with a large channel length). If the current through qn1 is less than that of mps1, transistor mps3 pulls the gate potential of mp3 low, thereby forcing current through qn3 and consequently through qn1. Transistor mps3 turns off when the current through qn1 is greater than the current through mps1, which occurs when the circuit is operating properly (characteristic of a bi-stable circuit).

Transistors qn1 and qn2 along with resistor R form a PTAT current generator. The unit size for the NPN transistors is chosen to occupy

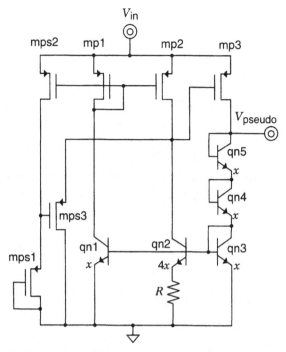

Figure 4.9   Preregulator circuit for generating a pseudo-supply voltage.

minimum silicon area. Since the accuracy of the current is not intrinsic, the benefits of using a larger geometry are not desired. As such, all NPN devices are minimum size except for qn2, which needs a larger size to produce a PTAT current. The size of qn2 is therefore designed to be four times larger than that of qn1 (unit size). A relatively small area spread between qn1 and qn2 does not require significant silicon area. The currents flowing through qn1 and qn2 are arbitrarily chosen to be 2.5 μA (mp1 and mp2 comprise a one-to-one mirror, thereby forcing the currents to be equal). A lower magnitude could have been chosen but, for the sake of noise immunity, it was not. At lower quiescent-current flow, the circuit is more vulnerable to reaching its off state when transient noise is injected through the substrate. As a result, resistor R is designed to be 14.3 kΩ,

$$I_{c2} = \frac{V_T}{R} \ln 4 \equiv 2.5 \ \mu A$$

or

$$R = \frac{V_T}{2.5 \ \mu\text{A}} \ln 4 = 14.3 \ \text{k}\Omega,$$

where I_{c2} is the collector current of qn2 and V_T is the thermal voltage. Transistors mp1, mp2, and mp3 are designed to be good current mirrors of each other, exhibiting reduced channel-length modulation effects. The source-drain voltages of both devices are equal. Furthermore, their channel length is chosen to be 10 μm (well above the minimum channel length allowed). The width of these devices, however, is designed according to the minimum input voltage requirement of the circuit (i.e., 1.5 V),

$$V_{in\text{-}min} = V_{SG1} + V_{CE1\text{-}min} = |V_{t1}| + V_{SD1\text{-}sat} + V_{CE1\text{-}min} \leq 1.5 \ \text{V}$$

thus,

$$V_{SD1\text{-}sat} \leq (1.5 \ \text{V}) - V_{CE1\text{-}min} - |V_{t1}|$$

$$\equiv \sqrt{\frac{2I_{c2}}{K'(W/L)_{mp1}}} \leq 0.5 \ \text{V}$$

or

$$(W/L)_{mp1} \geq 1.4,$$

where V_{SG1} is the source-gate voltage of mp1, $V_{CE1\text{-}min}$ is the minimum voltage across the collector and the emitter of qn1 (assumed to be 300 mV), $|V_{t1}|$ is the threshold voltage of mp1 (assumed to have a value of 0.7 V), K' is the characteristic transconductance parameter of the device (assumed to be 15 μA/V^2), and $(W/L)_{mp1}$ is the aspect ratio of mp1. Consequently, a channel width of 20 μm (twice its length) is chosen for mp1, mp2, and mps2. Transistor mp3 is designed to source the load current with a low source-drain voltage drop. This criterion is designed to cater to the minimum input voltage overhead of the circuit,

$$V_{in\text{-}min} = V_{pseudo} + V_{SD3\text{-}min} = (1.3 \ \text{V}) + V_{SD3\text{-}sat} \leq 1.5 \ \text{V}$$

thus,

$$V_{SD3\text{-}sat} = \sqrt{\frac{2I_{\text{Load}}}{K'(W/L)_{mp3}}} \leq 0.2 \ \text{V}$$

or

$$(W/L)_{mp3} \geq 150,$$

where I_{Load} is assumed to be less than 45 μA (40 μA demanded by the reference circuit and less than 5 μA allocated for transistors qn3 through qn4). The channel length for mp3 is chosen to be 4 μm, which corresponds to approximately 2 times larger than the minimum channel length for a 2 μm process. A minimum channel length is not chosen to alleviate the effects of channel-length modulation on the output resistance, which reduces the PSR performance of the circuit. The tradeoff, however, is silicon area and that is why the physical channel length is only twice the minimum allowed. Finally, transistor mps3 is chosen to be minimum size to reduce area overhead and transistor mps1 is designed to sink less than the minimum current flowing through qn2, which occurs at low temperatures and at high resistive values (typical tolerance is $\pm 20\%$).

$$I_{SD\text{-}mps1} \leq I_{2\text{-min}}^{c} \equiv \frac{V_T|_{\text{Temperature} = -40\ ^\circ\text{C}}}{R|_{+20\%}} \ln 4 \approx 1.6\ \mu\text{A}.$$

Due to the inherent variability of the PMOS device itself (up to 100% across process and temperature), its source-drain current ($I_{SD\text{-}mps1}$) is chosen to be less than 0.8 μA at room temperature, thus

$$I_{SD\text{-}mps1} = \frac{K'}{2}(W/L)_{mps1}(V_{SG} - |V_t|)^2 \leq 0.8\ \mu\text{A}$$

or

$$(W/L)_{mps1} \leq \frac{2(0.8\ \mu\text{A})}{K'(V_{SG} - |V_t|)^2}$$

$$\approx \frac{2(0.8\ \mu\text{A})}{K'(V_{\text{in-max}} - |V_t|)^2} \approx \frac{1}{810},$$

where $V_{in\text{-max}}$ is 10 V.

4.4 SUMMARY

Precision references are not merely high-order temperature-compensated circuits. They are circuits that reliably and predictably produce a reference voltage despite process, load, line, and transient variations. High overall accuracy is an exceedingly demanded commodity. Though not all circuits require it, more and more complex systems are dependent on it. Process-dependent variations impacting the performance include resistor matching and tolerance, transistor matching and tolerance, channel-length modulation, Early voltage, and so on. They alter the biasing condition for which the lowest TC is exhibited. Other factors that alter the operating performance of the reference include load and line regulation, not to mention PSR. Load regulation is considered when designing the output stage of the reference. Line regulation, on the other hand, is taken into account throughout the design of the overall circuit—input and output stages. Ultimately, to reduce the output sensitivity to the load, a regulated output may be designed. To decrease supply sensitivity, on the other hand, a preregulated supply is quite effective.

At this point in the text, most of the issues involved behind the design of references and bandgap circuits have been addressed. The temperature-compensated circuits of Chapter 3 are enhanced and complemented with the circuits and techniques shown in this chapter, which allow them to be relatively insensitive to load and input supplies. Depending on the demands of the system and the characteristics of the process, different circuit topologies within the set presented may be warranted. For instance, the output may be voltage, current, or mixed mode. It could also be regulated or not regulated. Similarly, the circuit, throughout, may adopt the use of cascodes or, perhaps, the use of a pseudo-supply. The next chapter deals with system-related issues such as initial accuracy (for trimming), package shifts, substrate noise, layout issues, characterization, and so forth. These topics are essential to the ultimate performance and utility of the reference!

APPENDIX A.4   ERROR SOURCES IN A TYPICAL FIRST-ORDER BANDGAP

The intrinsic characteristics of the bandgap reference are the temperature dependence of each major component, i.e., base-emitter voltage, proportional-to-absolute-temperature (PTAT) voltage, and curvature-correction voltage (for second-order references). These voltages,

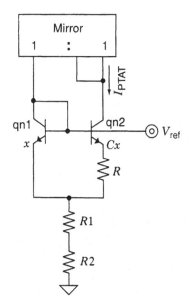

Figure A.4.1 Typical first-order bandgap circuit.

however, exhibit nonideal relationships to absolute voltage and to temperature. These nonidealities occur because of the parasitic effects involved in integrated circuit design. Most of the errors lie in the basic core of the reference (i.e., the base-emitter voltage used and the PTAT generator circuit). Figure A.4.1 illustrates a typical implementation of a first-order bandgap circuit. This reference is used to scrutinize the parasitic components that alter its ideal behavior. The applicability of this analysis can be extended to a wide range of reference circuits due to the recurrence of the same principles. The basic components are present in many, if not most, references.

The relationship of the reference is described by

$$V_{\text{ref}} = V_{BE1} + 2I_{\text{PTAT}}(R1 + R2), \qquad (A.4.1)$$

where V_{BE1} is the base-emitter voltage of transistor qn1 and the PTAT current (I_{PTAT}) is

$$I_{\text{PTAT}} \equiv I_{C2} = \frac{V_T}{R} \ln\left(\frac{Cx}{x} \cdot \frac{I_{C1}}{I_{C2}}\right), \qquad (A.4.2)$$

where I_{C1} and I_{C2} are the collector currents of transistors qn1 and qn2, respectively. The dependence of the base-emitter voltage to collector current I_C is

$$V_{BE} = V_T \ln\left(\frac{I_C}{J_S \text{Area}}\right), \qquad (A.4.3)$$

where V_T is the thermal voltage and J_S is the saturation current density.

RESISTOR MISMATCH

Resistor mismatch affects only the PTAT term of the reference as observed in equations (A.4.1) and (A.4.2), when combined. The mismatch can be mathematically described by

$$(R1 + R2)_x = \frac{R1 + R2}{R} R(1 + \delta_{RR}) = (R1 + R2)(1 + \delta_{RR}),$$

$$(A.4.4)$$

where $(R1 + R2)_x$ is the value of sum $(R1 + R2)$, taking the relative percent mismatch δ_{RR} to resistor R into consideration $((R1 + R2) \div R$ is the ideal gain-ratio $A)$. The ratio that allows the circuit to exhibit its lowest temperature drift over the specified temperature range is the ideal gain-ratio. Consequently, the relationship of the reference is affected and its deviation from its nominal value can be derived by

$$V_{\text{ref-}x} = V_{\text{ref}} + (V_{\text{ref-}x} - V_{\text{ref}}), \qquad (A.4.5)$$

where $V_{\text{ref-}x}$ is the value of the reference voltage when resistor mismatches are taken into consideration. By using equations (A.4.1), (A.4.2), and (A.4.4) in (A.4.5), the parasitic behavior of the resistor mismatch can be quantified,

$$V_{\text{ref-}x} = V_{\text{ref}} + 2I_{\text{PTAT}}(R1 + R2)\delta_{RR}. \qquad (A.4.6)$$

As a result, resistor mismatch not only changes the absolute value of the reference but it also adds a PTAT temperature component. This mismatch is typically addressed by trimming the circuit, particularly, by trimming $R2$.

RESISTOR TOLERANCE

The lot-to-lot variation of the absolute magnitude of the resistors also affects the reference. The variation of resistor R with respect to itself is defined as the resistor tolerance. The variation of sum $(R1 + R2)$ is absorbed by resistor mismatch since its variation is gauged against resistor R. This tolerance can be quantitatively described by

$$R_x = R(1 + \delta_{RA}), \qquad (A.4.7)$$

where R_x is the value of resistor R when absolute resistor tolerance δ_{RA} is taken into consideration. This variation affects the absolute value of the PTAT current, which ultimately affects the absolute value of the base-emitter voltage. This consequence is observed by deriving the effect of tolerance on the value of the base-emitter voltage, i.e.,

$$V_{BE1-x} = V_{BE1} + (V_{BE1-x} - V_{BE1}), \qquad (A.4.8)$$

where V_{BE1-x} is the value of the base-emitter voltage of transistor qn1 when resistor tolerance δ_{RA} is taken into consideration. Thus, equations (A.4.2), (A.4.3), and (A.4.7) are used in (A.4.8) to yield

$$V_{BE1-x} = V_{BE1} + V_T \ln\left[\frac{R}{R(1 + \delta_{RA})}\right] = V_{BE1} - V_T \ln(1 + \delta_{RA}).$$

$$(A.4.9)$$

As a result, the absolute tolerance adds a CTAT term behavior to the temperature drift of the base-emitter voltage and, ultimately, to the reference voltage, i.e.,

$$V_{\text{ref-}x} = V_{\text{ref}} - V_T \ln(1 + \delta_{RA}) \approx V_{\text{ref}} - V_T \delta_{RA}. \qquad (A.4.10)$$

The approximation only applies if δ_{RA} is much lower than 1, which is a reasonable assumption when considering that typical resistor mismatches are below 5%.

CURRENT MIRROR MISMATCH

A mismatch in the currents flowing through the collectors of transistors qn1 and qn2 affects the base-emitter voltage as well as the PTAT current and, ultimately, the PTAT voltage. The following relation can quantify the percent mismatch of the mirror:

$$I_{C1} = I_{C2}(1 + \delta_M), \tag{A.4.11}$$

where δ_M is the percent current mismatch between the collector currents in transistors qn1 and qn2. This mismatch affects the base-emitter voltage as well as the PTAT voltage term. First, reanalyzing the ΔV_{BE} loop derives the erroneous PTAT current (I_{PTAT-x}),

$$\bullet \quad I_{PTAT-x} = \frac{V_T}{R} \ln\left[\frac{I_{C2}(1 + \delta_M)C}{I_{C2}}\right] = \frac{V_T}{R} \ln[(1 + \delta_m)C]$$

$$= \frac{V_T}{R}\left[1 + \frac{\ln(1 + \delta_M)}{\ln C}\right] \ln C. \tag{A.4.12}$$

The effects of this current on the base-emitter voltage are observed by substituting equation (A.4.3) into (A.4.8) and using the current mismatch relationship in equation (A.4.11) as well as the erroneous current derived in equation (A.4.12),

$$V_{BE1-x} = V_{BE1} + V_T \ln\left[\frac{I_{PTAT-x}}{I_{PTAT}}\right] = V_{BE1} + V_T \ln\left[\frac{\ln[C(1 + \delta_M)]}{\ln C}\right]$$

$$= V_{BE1} + V_T \ln\left[\frac{\ln C + \ln(1 + \delta_M)}{\ln C}\right]$$

$$\approx V_{BE1} + V_T \ln\left[1 + \frac{\delta_M}{\ln C}\right] \approx V_{BE1} + V_T \frac{\delta_M}{\ln C}. \tag{A.4.13}$$

The final approximation is true only if δ_M and ($\delta_M \div \ln C$) are much less than 1, which is reasonable since current mismatch error is typically less than 10 to 15% and C is not usually larger than 12. The resulting effect on V_{BE} is similar to that of the absolute resistor tolerance. The error term propagates to the reference relationship in addition to a PTAT error term. Equation (A.4.5) is used in conjunction with (A.4.1) and (A.4.13) to elucidate the total effects of current mismatch on the reference voltage,

$$V_{\text{ref-}x} = V_{\text{ref}} + V_T \ln\left(1 + \frac{\ln(1 + \delta_M)}{\ln C}\right) + I_{\text{PTAT-}x}\delta_M(R1 + R2)$$

$$\approx V_{\text{ref}} + V_T\frac{\delta_M}{\ln C} + I_{\text{PTAT}}\left(1 + \frac{\delta_M}{\ln C}\right)\delta_M(R1 + R2), \quad \text{(A.4.14)}$$

where I_{PTAT} is the current flowing through the collector of transistor qn1 and the same approximations as for equation (A.4.13) are true.

TRANSISTOR MISMATCH

The area ratio between transistors qn1 and qn2 also exhibits some error. This error is termed *mismatch* (δ_{NPN}) and it manifests itself by yielding an effective gain-ratio of $(1 + \delta_{NPN})Cx$; in other words, the gain-ratio between qn1 and qn2 is $x:(1 + \delta_{NPN})Cx$. This also translates to an equivalent offset voltage,

$$V_{os} = V_{BE\text{-}x} - V_{BE} = V_T \ln\left[\frac{(1 + \delta_{NPN})C}{C}\right] \approx V_T\delta_{NPN}. \quad \text{(A.4.15)}$$

The mismatch between the two transistors in the circuit at hand affects the absolute value of the PTAT current by forcing an additional offset voltage across resistor R. As a result of I_{PTAT} changing, the base-emitter voltage is affected as well as the PTAT term of the reference relationship. By utilizing equations (A.4.2), (A.4.3), and (A.4.8), the effects of transistor mismatch (offset voltage V_{os}) can be derived, i.e.,

$$V_{BE1\text{-}x} = V_{BE1} + V_T \ln\left(\frac{I_{\text{PTAT}} + \dfrac{V_{os}}{R}}{I_{\text{PTAT}}}\right)$$

$$= V_{BE1} + V_T \ln\left(1 + \frac{V_{os}}{V_T \ln C}\right). \quad \text{(A.4.16)}$$

The final relationship of the reference also shows this error term in addition to an offset voltage term. The additional term is simply the offset voltage magnified by the gain-ratio between sum $(R1 + R2)$ and R. The effect is illustrated by using equations (A.4.1), (A.4.5), (A.4.15), and (A.4.16),

$$V_{\text{ref-}x} = V_{\text{ref}} + V_T \ln\left(1 + \frac{V_{os}}{V_T \ln C}\right) + 2\left(\frac{V_{os}}{R}\right)(R1 + R2)$$

$$\approx V_{\text{ref}} + V_T \ln\left(1 + \frac{\delta_{NPN}}{\ln C}\right) + 2\left(\frac{V_T \delta_{NPN}}{R}\right)(R1 + R2)$$

$$\approx V_{\text{ref}} + V_T \frac{\delta_{NPN}}{\ln C} + \frac{2 V_T \delta_{NPN}}{R}(R1 + R2), \qquad (A.4.17)$$

where I_{PTAT} has been replaced by $(I_{\text{PTAT}} + V_{os} \div R)$.

EARLY VOLTAGE

Early voltage errors associated with transistors qn1 and qn2 affect the reference by altering the PTAT current. Consequently, the base-emitter voltage relationship is also affected as well as the PTAT term of the reference. The effects on the PTAT current are observed by taking into account Early voltage on the base-emitter voltage relationship,

$$I_C = J_S \text{Area}\left(1 + \frac{V_{CE}}{V_A}\right)\exp\left(\frac{V_{BE}}{V_T}\right) \qquad (A.4.18)$$

or

$$V_{BE} = V_T \ln\left[\frac{I_C}{J_S \text{Area}\left(1 + \frac{V_{CE}}{V_A}\right)}\right], \qquad (A.4.19)$$

where V_A is the Early voltage. The PTAT current can now be reevaluated by using the above relationship for the base-emitter voltages of transistors qn1 and qn2,

$$I_{\text{PTAT-}x} = \frac{V_T}{R} \ln \left[\frac{C\left(1 + \dfrac{V_{CE2}}{V_A}\right)}{\left(1 + \dfrac{V_{CE1}}{V_A}\right)} \right], \qquad (A.4.20)$$

where V_{CE1} and V_{CE2} are the collector-emitter voltages of transistors qn1 and qn2, respectively. The effect on the base-emitter voltage is observed by using equation (A.4.8):

$$V_{BE1\text{-}x} = V_{BE1} + V_T \ln \left(\frac{I_{\text{PTAT-}x}}{I_{\text{PTAT}}} \right). \qquad (A.4.21)$$

This error propagates to the reference relationship in addition to also altering the PTAT term as demonstrated by

$$V_{\text{ref-}x} = V_{\text{ref}} + V_T \ln \frac{I_{\text{PTAT-}x}}{I_{\text{PTAT}}} + 2\frac{V_T}{R}(R1 + R2) \ln \left[\frac{\left(1 + \dfrac{V_{CE2}}{V_A}\right)}{\left(1 + \dfrac{V_{CE1}}{V_A}\right)} \right],$$

$$(A.4.22)$$

where equations (A.4.1), (A.4.5), (A.4.20), and (A.4.21) have been applied.

RESISTORS' TEMPERATURE COEFFICIENT

The temperature coefficient of the resistors also affects the temperature-drift performance of the reference. Resistors R, $R1$, and $R2$ should be made of the same material so that their TCs track. As a result, the PTAT term of the reference relation is unaffected ($V_{\text{PTAT}} \propto (R1 + R2) \div R$). The parasitic effects due to the TC of resistor R in the PTAT generator does, however, affect the reference. The following relation describes the behavior of a realistic resistor:

$$R(T) \approx R(T_r)\left[1 + A(T - T_r) + B(T - T_r)^2\right], \qquad (A.4.23)$$

where A and B are the linear and quadratic temperature coefficients, $R(T)$ is the resistance at temperature T, and $R(T_r)$ is the resistance at a reference or room temperature (ideal resistance). Using equations (A.4.2), (A.4.8), and (A.4.23) can thus derive the dependence of the base-emitter voltage on resistor R,

$$
\begin{aligned}
V_{BE1\text{-}x} &= V_{BE1} + V_T \ln\left[\frac{R(T_r)}{R(T)}\right] \\
&= V_{BE1} - V_T \ln\left[1 + A(T - T_r) + B(T - T_r)^2\right].
\end{aligned} \quad \text{(A.4.24)}
$$

Consequently, the error term is also present in the relationship of the reference, i.e.,

$$
V_{\text{ref-}x} = V_{\text{ref}} - V_T \ln\left[1 + A(T - T_r) + B(T - T_r)^2\right]. \quad \text{(A.4.25)}
$$

APPENDIX B.4   EFFECT OF THE RESISTORS' TEMPERATURE COEFFICIENT ON A REFERENCE WITH A CURRENT-MODE OUTPUT STAGE

In a current-mode output stage reference, the resistors' TC does not degrade the performance of the circuit relative to a voltage-mode topology. The TC of the resistors is basically canceled and the only effects present are those that also affect the voltage-mode topology (i.e., the TC of the base-emitter voltage as well as the PTAT voltage change, as discussed in Appendix A.4 and Section 4.1). Figure B.4.1 exemplifies a current-mode output structure. The relationship of the

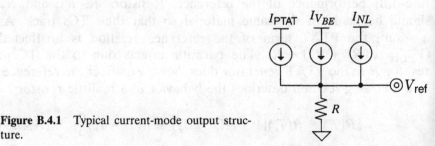

Figure B.4.1 Typical current-mode output structure.

reference voltage for this circuit is

$$V_{\text{ref}} = \left(I_{V_{BE}} + I_{\text{PTAT}} + I_{NL}\right)R \equiv \left(K_a \frac{V_{BE}}{R_a} + K_b \frac{V_T}{R_b} + I_{NL}\right)R, \quad \text{(B.4.1)}$$

where $I_{V_{BE}}$, I_{PTAT}, and I_{NL} are base-emitter voltage derived, PTAT, and nonlinear currents, respectively, while K_a and K_b are temperature-independent constants, V_{BE} is the base-emitter voltage, V_T is the thermal voltage, and R_a, R_b, and R are resistors fabricated with the same material. The substitutions made in equation (B.4.1) represent how the currents are, for the most part, physically defined. Since the temperature behavior of current I_{NL} is nonlinear in nature, the TC of resistor R simply modifies an already nonlinear voltage component. Hence, the TC is assumed to have an insignificant effect on the potential performance of the reference. Additionally, if the nonlinear current is derived from a voltage, then its dependence to a resistor is similar to $I_{V_{BE}}$ and I_{PTAT} and the same claim of TC cancellation applies ($I_{NL} = V_{NL} \div R_c$).

The temperature dependence of the resistors can be described as

$$R(T) = R(T_r)\left[1 + A(T - T_r) + B(T - T_r)^2\right], \quad \text{(B.4.2)}$$

where T is temperature, A and B are the linear and the quadratic temperature coefficients, and $R(T_r)$ is the resistance at room temperature T_r. Consequently, the reference voltage is reexpressed as

$$V_{\text{ref}} = \left[K_a \frac{V_{BE}}{R_a(T_r)} + K_b \frac{V_T}{R_b(T_r)} + I_{NL}\right]$$

$$\times \left[\frac{R(T_r)\left[1 + A(T - T_r) + B(T - T_r)^2\right]}{\left[1 + A(T - T_r) + B(T - T_r)^2\right]}\right] \quad \text{(B.4.3)}$$

or

$$V_{\text{ref}} = \left[K_a \frac{V_{BE}}{R_a(T_r)} + K_b \frac{V_T}{R_b(T_r)} + I_{NL}\right]R(T_r). \quad \text{(B.4.4)}$$

Hence, the temperature coefficients of the resistors are canceled as long as R_a, R_b, and R are all the same type (i.e., poly2 resistors).

BIBLIOGRAPHY

[1]   A.S. Sedra and K.C. Smith, *Microelectronic Circuits*. New York: Holt, Rinehart and Winston, 1987.

[2]   *Practical Design Techniques for Power and Thermal Management*. Analog Devices, 1998.

[3]   K.M. Tham and K. Nagaraj, "A Low Supply Voltage, High PSRR Voltage Reference in CMOS Process," *IEEE Journal of Solid-State Circuits*, vol. 30, no. 5, pp. 586–590, May 1995.

[4]   G.A. Rincon-Mora and P.E. Allen, "A 1.1 V Current-Mode and Piece-wise-Linear Curvature Corrected Bandgap Reference," *IEEE Journal of Solid-State Circuits*, vol. 33, no. 10, pp. 1551–1554, October 1998.

[5]   G.A. Rincon-Mora, *Current Efficient, Low Voltage, Low Drop-Out Regulators*. Ph.D. Dissertation. UMI, 1996.

CHAPTER 5

CONSIDERING THE SYSTEM AND THE WORKING ENVIRONMENT

Process variations and gradients across the die present significant challenges to the designer. These design obstacles have an impact on the design of both the circuit and the layout, especially in reference circuits. For instance, if the expected variation of the reference due to process is greater than the acceptable tolerance demanded by the system, which is a typical situation, a postfabrication trim network must be designed. The layout must therefore consider this characteristic in its allocation of space as well as in its dissipation of current, during trim routines. It must also minimize the full-scale range requirements of the trim network (maximum all-bit weight) by matching all key devices. More trim bits require more die area and longer test times, which increases overall cost. Process characteristics also change when encapsulating the reference circuit in a package. This phenomenon, of course, reduces predictability and induces offsets.

The system itself can also impose severe limitations on the reference. Many integrated circuits today, in fact, are geared toward the mobile market, which demands system integration, low quiescent-current flow, and low voltage. The implications of system integration are twofold: (1) the reference must use a minimum amount of silicon space and (2) the circuit must be insensitive to noise and various operating conditions. The latter is somewhat in conflict with low quiescent-current flow. Ultimately, the task of designing the integrated systems of today and tomorrow is difficult. This trait is especially true because, under a low-voltage environment, circuit techniques are greatly limited. Unfortunately, a harsh electrical environment also demands careful layout to reduce the effects of noise events as well as transient loads.

In the end, the utility of the reference as a whole depends on its performance with respect to the different factors that alter it. As such, characterization is an intrinsic part of the entire product-development process. The individual effects of each element affecting and/or describing the reference must be carefully ascertained to properly illustrate the overall impact of the reference on the system. The consequences must be independently rationalized, as a result. A well-identified set of specifications is especially important during the design phase of the system, when the performance of the reference needs to be anticipated. Once designed, though, the ability to predict the behavior of the reference is specified by the characterization parameters and they are only as valid as the process used to extract the parameters themselves. Consequently, a meticulous test environment and a well-planned test strategy are important features of any characterization process. After addressing the issues surrounding trim, package shifts, and system-induced requirements, this chapter will end with a discussion on characterization, the final gate before a product is released.

5.1 DESIGN OF THE TRIM NETWORK

Integrated references inherently suffer from practical nonidealities associated with the fabrication of circuits in any process technology. These parasitic effects manifest themselves in the form of current mirror mismatches, resistors' tolerance, resistors' temperature co-efficient (TC), Early voltage, channel-length modulation, resistor mismatches, transistor mismatches, package shifts, and input offset voltages. Some of these errors are systematic and effectively predicted by the simulator, like TC, Early voltage, and channel-length modulation. However, device mismatches, tolerance, and package effects are random in nature and will therefore vary from chip to chip, wafer to wafer, and lot to lot. The ultimate effects of these errors on the reference become increasingly critical as the order of temperature correction is elevated. Consequently, the robustness of a particular design depends on its susceptibility to process and package effects. In the end, all precision references are subject to these random parasitic effects and therefore require postfabrication adjustments (trim). The extent of the fine-tuning process depends on the particular circuit, the desired trimming accuracy, the resulting temperature-drift performance of the reference, and the package-induced shifts (related to mechanical stresses).

Since trimming is performed after the silicon has been processed, there is not a great deal of flexibility. It is usually done at the wafer level and seldom at the postpackage stage. The physical restrictions imposed on the trimming of postprocessed units are severe; neither additional devices nor reconnections are available. Additionally, the cost (time being a major contributing factor) associated with any trim network also limits the flexibility of the algorithm. As a result, typical networks are comprised of resistors that can be either short-circuited or open-circuited at room temperature. Trimming throughout the temperature range is possible but the time overhead is usually unacceptable for most commercial products. This restriction may not be an issue for military applications where product yield (percentage of devices meeting all specifications within a given process lot) can be low. The effectiveness of the trim network depends on its prefabrication design. What resistor is trimmed and how it is done can *make* or *break* an integrated reference. For instance, a second-order reference circuit is rendered useless if a trim is correcting a proportional-to-absolute temperature (PTAT) voltage error but the actual trim voltage itself is not PTAT in nature.

5.1.1 Trim Range

One of the most important aspects of trimming a precision reference is the TC of the error voltage causing the offset; the TC of the trim voltage must be equal to it. For the case of bandgap references, the errors are predominantly linearly dependent to temperature, as shown in section 4.1 of Chapter 4. The random offsets associated with the nonlinear correction components of second-order and higher-order references are typically small relative to the linear errors. This characteristic results because the magnitude of the nonlinearly correcting component is lower than the zero-order and first-order components, as depicted by V_{NL} in Figure 5.1. What may occur, however, is that a significant nonlinear systematic offset may be present. The systematic error may result from using poor electrical models during the design phase. This type of offset is not trimmed but instead "centered" based on experimental characterization of the circuit in the given process

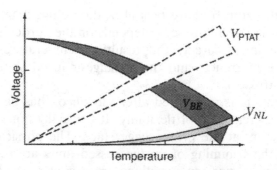

Figure 5.1 Variations of the different temperature-dependent components of a bandgap reference.

technology. Some of the more prevalent errors that cause variation in the PTAT voltage are resistor mismatch, NPN transistor mismatch, and current mirror mismatch. Similarly, mirror mismatches, resistor tolerances, and saturation current tolerances cause variations in the base-emitter voltage. The initial tolerance of the overall bandgap reference voltage is between ± 2 and 3.5%. The derived tolerance of the bandgap circuit used in Example 4.1 of Chapter 4 showed a variation of approximately $\pm 2.1\%$ (± 25.2 mV $\div 1.2$ V).

If a linear offset with respect to temperature is assumed, then trimming at a single temperature point is all that is necessary to effectively cancel the error [1]. Theoretically, the values corresponding to two temperature settings are needed to effectively compensate a first-order error. However, the value of the components and therefore the reference itself are already predetermined at 0 °K. The PTAT and the nonlinear voltages are zero and the base-emitter voltage is V_{go} (process-dependent constant) at 0 °K. Consequently, the linear error is effectively canceled by adding or subtracting a linearly correcting component to the reference at any single temperature setting (typically room temperature),

$$V_{ref} = V_{ref-i} \pm V_{error-T_r} \approx V_{ref-i} \pm e_{linear} T|_{T_r}, \qquad (5.1)$$

where V_{ref-i} is the ideal reference voltage, $V_{error-T_r}$ is the offset error at reference temperature T_r, and e_{linear} is the corresponding linear coefficient of $V_{error-T_r}$. Thus, a linear error is canceled if a linearly dependent trim voltage (V_{trim-T_r}) equal to $V_{error-T_r}$, at any reference temperature setting is added,

$$V_{ref} = V_{ref\text{-}i} \pm V_{error\text{-}T_r} \mp V_{trim\text{-}T_r} \approx V_{ref\text{-}i} + (\pm e_{linear} \mp k_{trim})T|_{T_r}, \quad (5.2)$$

where k_{trim} is the variable trimming coefficient.

If significant higher-order errors exist, more temperature settings may be required to effectively trim the reference. Appendix A.5 shows a trimming algorithm that can essentially trim any three linearly and/or nonlinearly temperature-dependent currents in a current/voltage mode output stage. Data obtained throughout the temperature range is used to extrapolate the behavior of the individual components. As a result, the coefficients of each component can ultimately be manipulated to minimize the overall temperature-drift performance. Unfortunately, this method is costly because it requires a significant amount of time, not to mention a thermal stream or an oven, for the trimming process. The assumption for most curvature-corrected bandgap references, however, is that the first-order offset error predominates the higher-order errors, thereby making a single temperature trim effective.

The design of the trim range is defined by the fundamental accuracy desired and by the expected tolerance of the untrimmed reference. In other words, the number of trim bits is determined by the least significant bit (LSB) value required and by the initial full-scale (FS) tolerance expected. For instance, a $\pm 0.5\%$ ($V_{\pm\%accuracy} = \pm 6$ mV) 1.2 V reference that exhibits a total initial tolerance of $\pm 2.5\%$ (half the full-scale voltage (V_{FS}) requirement, ± 30 mV) should have an LSB voltage (V_{LSB}) of at least 3 mV with five bits (# Bits) of trim (assuming a binarily weighted algorithm).

$$V_{LSB} \leq \frac{V_{\pm\%accuracy} V_{ref}}{K_c} \quad (5.3)$$

and

$$V_{FS} = V_{LSB}(2^{\#\,Bits} - 1)$$

or

$$\# \text{Bits} \geq \frac{\ln\left(\dfrac{V_{FS}}{V_{LSB}} + 1\right)}{\ln 2}, \quad (5.4)$$

where K_c is a "comfort factor" (two for this discussion). The comfort factor K_c, as the name implies, is an engineering index that balances cost with performance; both cost and performance increase as K_c is enlarged. The TC and the line regulation performance of the reference ultimately limit the desired tolerance of the circuit after trimming, which defines the value of the LSB. Since these are centered about an ideal reference voltage, any deviation in the actual value will affect the overall accuracy performance. For example, if the theoretical value at which the reference exhibits its lowest TC of 40 ppm/°C is 1.195 V and the trim network could only get 1.185 V, then the TC is most likely greater than the projected 40 ppm/°C. This reasoning explains why K_c is chosen to be a minimum of two. Greater values of K_c yield more accurate results but with diminishing returns and increased overhead. Ultimately, K_c is limited by available silicon area for placing probe pads and by the cost of increasing the number of trim bits. Systematic package-induced offsets can also be anticipated and canceled during wafer-level trimming. This process, however, requires that package effects be well-characterized (magnitude and temperature coefficient). Typically, package shifts are lower than ± 6 mV. Though more complex in design, a postpackage trimming algorithm is more effective at compensating for package shifts. The temperature dependence of the shift still requires scrutiny, though.

Theoretically, the reference voltage is centered about the trim code corresponding to half V_{FS}. In other words, the reference circuit is designed to generate the ideal voltage value (disregarding process variations) when a trim code of half V_{FS} is programmed (10000... or 01111...). This code results because the process errors are random, thereby potentially requiring equal positive and negative adjustments. Alternatively, the most significant bit (MSB) may be used as a *pole bit*, determining the polarity of the trim. In this latter case, the nominal value of the reference is centered at 00000... or equivalently at 10000..., where the first bit is the pole bit.

5.1.2 Trimming Techniques

At this point, the actual trimming mechanism is ready to be examined. There are three basic types of trimming techniques: Zener zap, fusible links, and laser trim. The process technology and the set of masks chosen for the particular design typically determine what trimming technique is to be adopted. Most process technologies today support Zener zap and fusible links. Laser trim, though more enticing from the design perspective, is more expensive and therefore less commercially attractive.

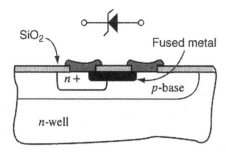

Figure 5.2 Cross-section of a zapped Zener diode.

Zener Zap   The technique of Zener zapping provides the capability of *short-circuiting* two nodes such as the two terminals of a resistor. Basically, a relatively large amount of current is forced to flow into the cathode (*n*-type material of a *p-n* junction diode) of a small Zener diode to short-circuit both terminals of the diode [2]. Typically, this current ranges from 200 to 300 mA requiring a large voltage drop across the cathode and anode (*p*-type material of a *p-n* junction diode) terminals of the diode. The large reverse current through the diode dissipates enough localized power to permanently destroy the *p-n* junction. The metal melts through the contact opening and is swept over the junction between the oxide layer and the silicon surface resulting in a short-circuit between the anode and the cathode of the device. Figure 5.2 illustrates the cross section of a zapped Zener diode [3]. Due to the large nature of the currents, special precautions are necessary to prevent permanent damages from occurring to the surrounding IC. Zener zapping is highly reliable and stable over time [1].

Fusible Links   Fuses, unlike Zener-zapping diodes, are normally *short-circuited* devices capable of being *open-circuited* once trimmed. The fuses are typically fabricated with aluminum or polysilicon. They are physically destroyed when a significant amount of current is forced to flow through them, thereby creating an open circuit. Approximately less than 5 to 6 V is necessary for this nonrecoverable mechanism to occur. Fusible links are therefore less intrusive than Zener-zapping diodes, which usually require higher voltages to program. Precautionary steps are still necessary to prevent metal regrowth resulting from "on-chip electromigration" [4].

Laser-Trimmable Resistors   Laser trim, like fusible links, can be used to cut metal links already short-circuiting resistor segments.

These breaks are more reliable and require less invasive methods than the fusible links. Another genre of laser trim physically alters the shape of thin-film resistors. As a result, the resistance of a single resistor is effectively modified. The resolution is not limited by the digital partitioning of resistors. Zener zapping and fusible links, unlike laser trim, are ultimately limited to the value of the least significant bit (LSB) to sum a trim voltage onto the reference. In other words, only incremental and finite resistor fragments are either short-circuited or open-circuited. Laser-trimmable resistors circumvent these limitations. There are no digital bits defining the range and accuracy of the trim. The equivalent resistance is adjusted to a wide range of values since the laser redefines the physical shape of the device. The resistance is continuously monitored while an L-shaped groove is carved into the resistor structure by a laser beam having a typical spot size of 1 μm [3]. Each resistor is aligned and trimmed individually. A cut perpendicular to the direction of the current is constructed first and then, as the resistance approaches its desired value, a parallel cut is made.

The overhead of laser trimming is time and money. The process requires expensive equipment and considerable time to align and to trim each resistor. Many processes today do not cater to such practices. However, as the demand for higher performance increases, they will soon become more common. Table 5.1 shows a qualitative comparison of the different characteristics of the three aforementioned trimming techniques. Overall, laser trimming is the most effective and the least power-intrusive method of trimming a resistor. However, Zener-zapping diodes and fusible links are the most economical. The nondigital adjustment category refers to the analog capability and nondiscrete nature of changing the value of a resistor, which laser-trimmable thin-film resistors display. "Stress on IC" refers to the power the die is forced to dissipate during the trimming process.

TABLE 5.1.. Qualitative Comparison of the Different Trimming Techniques

Trimming Techniques	Normally Open	Normally Closed	Nondigital Adjustment	Stress on IC	Cost
Zener zap	✔			High	Low
Fusible link		✔		Moderate	Low
Laser trim		✔	✔	Low	High

Design Example 5.1: Design a trim network for a second-order bandgap Brokaw cell with current-mode curvature correction (PTAT^2 current flows through a series resistor in the output stage) whose input power supply voltage is 5 V. Assume that the use of fusible links is adopted for trimming. Also assume that the reference voltage is expected to have an initial tolerance of $\pm 5\%$, that its target tolerance is $\pm 2\%$, and that the total PTAT and PTAT^2 resistance values required to nominally center the circuit are 130 and 21.5 kΩ, respectively.

Since most of the errors associated with a bandgap reference are PTAT, the trim voltage must therefore be PTAT in nature. As such, the PTAT component must be variable and programmable. Figure 5.3 shows the implementation of a PTAT trim voltage in a curvature-corrected bandgap reference. It is noted that resistor R_2 is not used as the variable resistor because its total current is not linear; it is the summation of a linear and a nonlinear current. In any case, the LSB voltage required is

$$V_{LSB} = \frac{V_{\pm\% \text{ accuracy}} V_{\text{ref}}}{K_c} \approx \frac{(0.02)(1.2\text{ V})}{2} = 12\text{ mV};$$

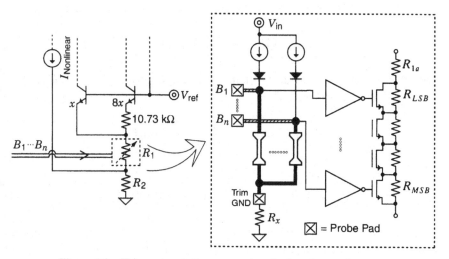

Figure 5.3 Trim network for a second-order bandgap reference.

thus, the resistance of the LSB resistor is

$$R_{LSB} \le \frac{V_{LSB}}{2I_{PTAT}} = \frac{12\ mV}{2\left(\dfrac{V_T \ln 8}{10.73\ k\Omega}\right)} \approx 1.2\ k\Omega,$$

where the reference voltage is approximated to 1.2 V and the comfort factor K_c is chosen to be two. The number of bits and, consequently, the number of binarily weighted resistors required in the network is four,

$$\#\ Bits \ge \frac{\ln\left(\dfrac{V_{FS}}{V_{LSB}} + 1\right)}{\ln 2} = \frac{\ln\left[\dfrac{k_c 2(5\%)V_{ref}}{V_{\pm\%\ accuracy}\ V_{ref}} + 1\right]}{\ln 2}$$

$$= \frac{\ln\left[\dfrac{2(0.1)}{0.02} + 1\right]}{\ln 2} = 3.5\ Bits.$$

Now, the reference voltage is nominally centered when the total resistance of R_1 and R_2 is 130 kΩ, which occurs when the trim resistor is half its full-scale value (all trim resistors are short-circuited except for R_{MSB} or equivalently only R_{MSB} is short-circuited. Thus, the nontrimmable resistor in the trim network (R_{1a}) is defined to be

$$R_{1a} \equiv (130\ k\Omega) - R_2 - R_{MSB},$$

where R_2 has been predefined in this design problem to be 21.5 kΩ (PTAT^2 resistance) and R_{MSB} is

$$R_{MSB} = 2^{(\#\ Bits - 1)} R_{LSB} = 9.6\ k\Omega;$$

thus,

$$R_{1a} \equiv (130 - 21.5 - 9.6)\ k\Omega = 98.9\ k\Omega.$$

One option for the physical implementation of the trim is to place the fuses across the terminals of each binarily weighted resistor. This approach is not adopted however! The current necessary to break the fuse is on the order of tens of milliamps (requiring approximately 5 to

6 V), thereby possibly changing some of the electrical characteristics of the resistors and the surrounding circuitry. Consequently, the fuses are digitally buffered with inverters and switches. The "on" resistance of the MOS switches short-circuiting the terminals of the trim resistors must be sufficiently low to prevent significant errors from altering the trim voltage. Therefore, $R_{DS\text{-on}}$ for these transistors is chosen to be less than 25% of the total resistance of R_{LSB},

$$R_{DS\text{-on}} \approx \frac{1}{K'(W/L)(V_{GS} - V_t)} \leq 0.25 R_{\text{LSB}}$$

or

$$(W/L) \geq \frac{1}{K'(V_{in} - V_t)(0.25 R_{LSB})} = 16,$$

thus choosing

$$(W/L) \equiv 30,$$

where the transconductance parameter K' and the threshold voltage V_t are assumed to be 50 $\mu A/V^2$ and 0.7 V, respectively, while the inverter driving the gate of each switch is assumed to be CMOS (V_{GS} can be raised to V_{in}). The aspect ratio is chosen to be roughly twice the minimum of 16 because the MOS resistance, over process and temperature, can vary up to approximately 100%.

The input voltages of the inverters must be either lowered or raised close to the supply rails to reliably define the gate voltage of the MOS switches. The fuses, when not blown, provide a low resistive path to ground. However, the input of the inverter must be forced to a high voltage when the fuse is blown. This task is achieved by forcing a small current to flow to the input node of the buffer inverter. A small voltage drop is created across the fuse when it is not blown but its potential is well below the threshold of the inverter (the resistance of the fuse is typically less than 200 Ω). When the fuse is blown, however, the current charges the capacitance of the input node to a few millivolts below the input power supply.

The process of breaking the fuse requires a considerable amount of current to flow through the fuse and back to ground. As a result, a few precautionary steps are taken to protect the small current source as well as to prevent current flow through the substrate of the IC. Toward this end, a diode is added in series to prevent current backflow through the current source when the fuse is programmed. A separate ground pad (trim ground) is also added to force the large current to flow through the pad and not through the ground of the IC. Resistor R_x, residing between trim ground and chip ground, is necessary to ensure that the fuse is pulled to the ground potential during normal post-trim operation. On one end, resistor R_x must be sufficiently low to guarantee that the input of the inverter is in a stable low state while, on the other hand, R_x must be large enough to prevent significant current to flow through chip ground during the trimming process.

Trim ground is externally short-circuited to chip ground during the trim, thereby impeding current from flowing through R_x. The assumption is that a low resistive path exists between trim ground and the common ground line of the instruments. The flow path of the large trim current is illustrated in Figure 5.3 by the thick lines (solid and hashed). The metal width of these traces must be large to reduce the equivalent series resistance and therefore reduce the power dissipated in the IC during the trimming procedure. The metal line must also abide by the current density requirements of the given process technology. For reliability, the voltage at the intersection of the solid and hashed thick lines must not exceed the breakdown voltage of the diodes added to protect the current sources. As a result, the resistance associated with the thick solid lines cannot be large. The resistance of the thick hashed lines simply increases the voltage overhead of the instrumentation that provides the trim current ($V_{instrument} = I_{trim}(R_{solid} + R_{hashed} + R_{fuse})$).

5.2   PACKAGE-SHIFT EFFECTS

As mentioned in the previous discussion, the process of packaging a reference circuit induces variations on the output voltage. Particularly, the mechanical stresses superimposed on the die alter the characteristics of the p-n junction, which, of course, is the basic core of most reference circuits [5]. Since the mechanical stresses vary from package to package, package-induced effects are dependent on package and vary accordingly. Ceramic packages, unlike the plastic counterparts, do not cause significant stresses on the die. Unfortunately, ceramic packages are not as cost effective, as compact, or as moisture insensitive as plastic packages, which do impose severe stresses on the die,

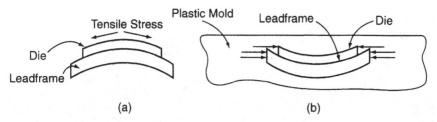

Figure 5.4 Mechanical stresses on the die: (a) tensile stress during the cool-down process of die attachment and (b) plastic-encapsulation effects.

thereby leading to parametric shifts, metal movement, and even die cracks [6, 7].

The two packaging steps mainly responsible for package-shift effects are die attachment and plastic encapsulation [8]. Differences in thermal expansion coefficients cause the die to undergo physical stresses. In the case of die attachment, once the die is physically attached to the leadframe, the die effectively bends as the die-leadframe combination is allowed to cool down to room temperature (the process of attachment takes place at elevated temperatures). This bending results because the die and the leadframe contract at different rates, thereby creating the tensile stresses shown in Figure 5.4(a). In particular, the leadframe, with a thermal expansion coefficient of roughly $16.5 \times 10^{-6}/°C$ for copper, contracts more than the silicon die, which has a lower thermal expansion coefficient (approximately $3.5 \times 10^{-6}/°C$). Different types of die-attach compounds can be used to mitigate these effects. Both ceramic and plastic packages suffer from this effect but their *p-n* junction repercussions are not greatly significant.

Plastic encapsulation, however, does impose significant stresses on the die. Its effects, as shown in Figure 5.4(b), counteract those of the attachment process. Most of the plastic molding is done at a temperature of 175 °C, to lower the viscosity of the plastic mold. The plastic, which has a typical coefficient of thermal expansion greater than ten times that of silicon, transmits an ever-increasing stress to the chip as the package cools from molding to ambient temperature. The plastic mold essentially contracts on the sides and on the top of the die, thereby bending it in a concave fashion. The resulting horizontal compressive die-surface stresses are much larger than the vertical stresses because the plastic lateral wall is much wider than the thin top layer of plastic [9, 10]. Vertical stresses increase with thicker layers

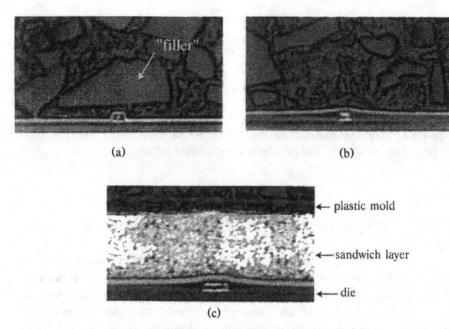

(a)   (b)

(c)

Figure 5.5 Cross-sectional images of dies: (a) nonplanarized, (b) planarized, and (c) planarized with an additional mechanically compliant layer ("sandwich layer").

of plastic mold. Compressive stresses, as a result, tend to be highest toward the center of the die. They also tend to be uniform and consistent from package to package. Shear stresses, on the other hand, are highest at the corners and edges and they are more statistically random in nature. What this means, with regard to the reference, is that the physical location of the layout, within the die, is important and it will affect how the package affects the reference.

One of the main vertical compressive stress-related failure mechanisms reported in the literature is the filler-induced mechanism [11]. The plastic mold consists of silica fillers that vary in size and shape, as shown in Figure 5.5(a). The fillers are used, among other reasons, to reduce the thermal coefficient expansion of the package to prevent destructive effects like corner and passivation layer cracking as well as metal-line shifts. Depending on the size, shape, and orientation of these fillers, they exert intense stress fields on localized regions of the die. The unpredictable nature of the location of these fillers is the main culprit for the packaged-induced variations in *p-n* junction characteristics.

The bandgap energy and the saturation current of *p-n* junction diodes change as a result of these stress-fields [5]. Package effects cause the minority carrier density concentration to increase, which causes an increase in saturation current, thereby decreasing the diode-voltage and ultimately reducing the bandgap reference voltage,

$$\Delta V_{BG} \approx V_T \ln\left(\frac{I_{s1o}}{I_{s1}}\right) \approx V_T \ln\left(\frac{1}{\gamma(\varepsilon)}\right) = -V_T \ln[\gamma(\varepsilon)], \quad (5.5)$$

where ΔV_{BG} is the variation of the bandgap voltage as a result of package shift, V_T is the thermal voltage, I_{so} and I_s are the unstressed and stressed saturation currents of the *p-n* junction, respectively, and γ is a function of the position-dependent strains (ε) in the junction. Appendix B.5 shows a detailed derivation and a more explicit explanation for the above-referenced shift in the reference voltage.

Theoretically, postpackage stress is a result of a difference in the thermal expansion coefficients of the plastic mold and the silicon die. Therefore, it can be shown, approximately, that:

$$\varepsilon \propto s \propto \left(T_{sp} - T\right), \qquad T < T_{sp} \tag{5.6}$$

where s is the stress on the die and T is temperature, which is below the molding set point T_{sp} (usually around 175 °C) [12]. The stress placed on the die by the plastic is directly proportional to the molding set temperature and varies with operating temperature. The end result is that the die becomes less stressed at higher temperatures (totally unstressed at molding set point temperature T_{sp}), thereby exhibiting an additional positive TC, which is actually parabolic in nature, as seen in Appendix B.5.

The effects of the differences in thermal expansion coefficients of the die, leadframe, and plastic mold on the bandgap reference while going through the die-attach and plastic-encapsulation process are global and, relatively speaking, fairly systematic in nature. They cause a decrease in reference voltage of roughly 3 mV at room temperature while yielding a positive TC of approximately 0.044 mV/°C, which varies with bond and mold compounds. To minimize this shift, as well as its randomness, the circuit should be placed toward the center of the die. Additionally, a full postpackaged characterization is required to fully ascertain the temperature-drift implications of the circuit and package. In other words, the TC of the package offset should be included in the optimization of the circuit when ascertaining a trim-target voltage, the "magical bandgap voltage." Consequently, establishing the reference voltage at room temperature for which the best TC occurs must be done after packaging.

Now, the statistical random component of the package drift is mainly the result of the silica fillers in the plastic, which exert intense and unpredictable localized fields of stress throughout the die. To address these variations, the addition of a stress-relief layer as thin as 15 μm has proven effective. Even planarization (Figure 5.5(b)) has shown improvements between 16 and 18%. When a mechanically compliant relief layer of 15 μm is added to an already planarized die (as shown in Figure 5.5(c)), an additional improvement of approximately 35% is also observed.

5.3 SYSTEM-RELATED ISSUES

There is currently a large and growing market demand for mobile battery-operated products, such as cellular phones, pagers, camera recorders, and laptops [13]. The thrust of the requirements is fueled by high performance under a low voltage and an environment characterized by low quiescent-current flow [14]. In fact, performance is many times optimized to maximize the efficiency and the longevity of the battery. Single nickel-cadmium (NiCd) and nickel-metal-hydride (NiMH) battery cells initially exhibit 1.5 to 1.6 V but they eventually collapse to approximately 0.9 V. The demand for low voltage is further corroborated by the breakdown voltage requirements of emerging technologies. As the packing densities increase, the breakdown voltages decrease, thereby imposing upper limits on power supply voltage [15, 16]. As a result, reference circuits that can operate at low input voltages and with low quiescent-current flow are in high demand! Additionally, single chip solutions, also in high demand, incorporate the generators of noise alongside sensitive analog circuitry like the reference. Consequently, high power supply rejection (PSR) and general high noise rejection are required.

5.3.1 Circuit Implications

The implications of low voltage, low quiescent-current flow and single IC solutions are severe. The uses of common techniques like emitter/source followers, Darlington configurations, cascoding devices, diode-connected level shifters, and so on are discouraged while the demands of PSR performance and accuracy are heightened. Current-sensitive paths (paths of current from input power supply to ground) must also be reduced to minimize quiescent-current flow. As a result, intelligent yet simple reference circuits with fewer electrical components are desired. Similarly, designs must address only the specific requirements demanded by the system to simplify circuit topology and therefore minimize quiescent-current flow. For instance, if the input supply voltage exhibits insignificant steady-state variations and low noise, neither preregulated pseudo-supplies nor cascoding devices, used to improve PSR, are necessary. The overhead in voltage and in current associated with a preregulator and/or series cascoding devices is circumvented.

Special care must be placed in designing the robustness of the circuit with respect to conditions that transiently cause the reference to collapse, like systematic noise injection. The reason for this requirement is that transient noise can cause the reference to transiently approach its zero-current state, thereby requiring a relatively fast startup circuit. For instance, a pulse-width-modulated (PWM) controller IC inherently generates noise in the substrate because of rapid voltage and current transitions. Since the substrate is shared, large transient noise impulses are coupled back into the reference circuit. This noise can momentarily pull internal critical nodes to ground or to the input supply. Noise is also injected through the supply voltage because of the innate large current transients of the system. The noise injected, the capacitance around sensitive nodes, and the speed with which the circuit pulls out of the zero-current state determine the susceptibility of the reference circuit to noise.

5.3.2 Layout Implications

Many of the system issues that degrade the performance and the quality of the reference come in the form of noise and thermal gradients. Systematic noise, like a clock or a switching regulator, may couple into the reference by means of *cross talk* and/or *substrate injection*. Consequently, the placement of paths whose signals have a frequency component must be carefully routed away and around the physical location of the reference. Generally, the reference must be distanced from the noise generators (e.g., the power drivers in a PWM controller). The space between the circuit and heat sources like power

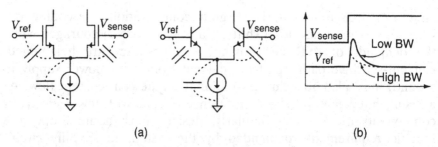

(a)   (b)

Figure 5.6 Noise coupling through the differential input pair of an amplifier.

output transistors, which carry large current densities, must also be maximized. The thermal gradients are greatest near these heat generators. Moreover, the reference should also be placed, as discussed earlier, toward the center of the die to minimize package-induced offsets resulting from mechanical stresses.

An analysis of the system with respect to noise generation is warranted to optimize layout and minimize injection. For instance, the fact that a metal trace carrying the reference signal physically ends in the input pair of a comparator or an amplifier does not necessarily imply that the path is immune to noise. If the other input of the comparator or amplifier has a noisy signal, the noise will couple through the input pair back into the reference. This phenomenon, sometimes referred to as *kickback noise*, is prevalent in switching regulators where the output ripple is sensed against a reference voltage through a comparator. The noise is coupled through the base-emitter/gate-source capacitance of the devices in the differential input pair, as illustrated in Figure 5.6. The extent of the noise reflected in the reference voltage is mitigated by the shunt capacitance between the reference and ground as well as by the bandwidth (speed) of the regulated reference. An unregulated reference will depend entirely on the shunt capacitor to minimize the effects of noise injection.

For the case of regulated references, the closed-loop bandwidth determines the speed with which the circuit will react to oppose the effects of noise. This is graphically illustrated by Figure 5.6(b). The time required for the reference to react determines the peak voltage variation of the reference voltage. A fast circuit reacts to oppose the transient noise before the voltage variation reaches undesired levels. As a result, the peak-to-peak voltage variation of the reference is reduced as the speed of the circuit increases, thereby improving effective overall accuracy. Increasing the shunt capacitance to ground also minimizes this variation; however, the stability (phase margin) of the feedback loop may be compromised. Settling time increases with degraded phase margin, of course. The repercussions of systematic noise being injected into the reference may manifest themselves as system instability. The output of a DC-DC converter, for instance, may exhibit unexpected frequency components as a result of a transiently varying reference.

Design Example 5.2: Design the architecture requirements of a $\pm5\%$ bandgap reference circuit with 40 dB of PSR (up to 1 kHz). Assume the temperature range is from 0 to 125 °C and a noisy supply voltage powers the circuit whose total steady-state variation is less than ±50 mV_{DC} in a battery-operated environment.

A first-order bandgap reference suffices for the aforementioned sample application. The temperature coefficient of a first-order circuit is between 20 and 100 ppm/°C, which yields ±1.5 to 7.5 mV of variation over the specified temperature range. A $\pm5\%$ reference can tolerate roughly ±60 mV of variation (assuming a bandgap voltage of 1.2 V), thereby leaving approximately ±52.5 mV for variations resulting from process tolerance and steady-state changes in supply voltage. The effects of the input supply varying ±50 mV_{DC} on the reference voltage are minimal. However, the effects of supply noise on the reference are sufficiently significant to warrant a design that caters to good PSR performance. As such, a pseudo-preregulated supply is chosen to be part of the design despite the line regulation performance required. A low-impedance output is desired to maximize PSR performance, thereby intimating a regulated output stage. Since a battery powers the system, low quiescent-current flow must be inherent to the solution. A voltage-mode output stage is simple, efficient, and reliable, requiring minimal quiescent current.

Ultimately, based on the choices already made, the reference voltage variation due to line regulation is going to be less than ±1 mV (±50 mV $\div$ PSR, where PSR is specified to be greater than 40 dB). As a result, a total variation of ±51.5 mV is allocated for process tolerance. This tolerable inaccuracy (equivalent to $\pm4.3\%$) along with the initial process-induced variation of the reference determines the trimming requirements of the circuit. Since typical bandgap references have an initial accuracy between ±2 to 3.5%, no trimming is required to achieve a $\pm5\%$ total variation! In summary, a first-order, untrimmed bandgap circuit with a preregulated pseudo-supply voltage and a regulated voltage-mode output stage is appropriate for the design problem stated in this example.

5.4 CHARACTERIZATION

Characterization is important for two purposes: (1) to ascertain if an adjustment is necessary (recenter the reference voltage according to experimental data) and (2) to document the performance and parametric limits of the circuit. Most precision references require some design adjustments when first tested in a new process or when first designed. Modifications are typically necessary because the models used in the initial design phase exhibit inaccuracies with respect to the real devices. Most commonly, the temperature-drift performance is the metric sought and evaluated. The other parameters describing the overall performance of the reference specify the operating limits of the particular design. The important aspects, requiring scrutiny, are line regulation, power supply rejection (PSR), "startup" robustness, quiescent current, and temperature coefficient (TC). Load regulation (output voltage variations resulting from changes in load current) typically refers to linear regulators as opposed to references. For the case of references, the load typically does not vary significantly. In any case, the load performance of the reference may still be evaluated, especially if its output impedance is low.

All measurements are validated through statistics; in other words, the specifications are considered valid only if a large portion of the units perform within the specified parametric window. The accuracy of the equipment used to gauge the parameters of the reference must exceed the accuracy of the circuit by at least three to four times. For example, a measurement system having 10 mV of resolution is not suitable to measure the TC of a 1% 1.2 V reference, which has a total variation of less than 12 mV. Furthermore, the noise injected to the reference, while testing, needs to be low. This requirement is especially important with references that exhibit high output impedance. A buffer may be necessary to isolate the sensitive reference from the noisy environment. The buffer, however, cannot load the reference with current, nor can it introduce a voltage offset to the measurement unless it is a well-characterized offset voltage whose value is taken into account when the data is analyzed. A chopper-stabilized buffer may be appropriate as long as the noise generated from the clock used in the buffer does not affect the measurement.

The temperature coefficient of the circuit is measured by placing the IC in an oven or a thermal stream and monitoring the reference voltage at different temperature settings. The biasing components and the experimental printed circuit board (PCB) need to be rated to tolerate the temperature extremes dictated by the specifications. Furthermore, the electrical biasing conditions must be consistent throughout the temperature range. For example, the input supply voltage (taken from a regulator onboard the same PCB) should exhibit low voltage variations throughout the temperature range. Ultimately, these tests must be done once the die has been packaged to include the mechanical-stress effects of the package itself.

The *soak time* (time allowed, after the temperature has settled, before collecting data) must be sufficiently long to permit the reference and the surrounding biasing components to settle to their steady-state condition. The minimum soak time is determined by increasing the soak time in an iterative fashion until no more differences in the value of the reference are encountered for a particular temperature setting (using the same unit device). As with most measurements, soak time must be established statistically to ensure that the data is reliable. Though often neglected, *thermal hysteresis* is also an important characteristic. If there is no difference in temperature drift performance when ascending as when descending through the temperature range, the circuit does not exhibit thermal hysteresis. If the soak time is not long enough, however, externally induced thermal hysteresis will be present, thereby nullifying the experiment. As a result, soak time and thermal hysteresis must be evaluated coherently.

The reference voltage is said to have thermal hysteresis only if hysteresis is still prevalent after soak time has been determined to be sufficiently long. For the case of true thermal hysteresis, the total temperature drift of the reference is defined by the voltage extremes measured throughout the ascending and the descending temperature sweeps. For instance, a reference voltage has a total temperature-induced variation of 10 mV if the reference peaks at 1.184 and 1.189 V when the temperature is swept in the negative direction and at 1.179 and 1.187 V when the temperature is swept in the opposite direction. Unfortunately, the factory test floor cannot afford to spend all the time necessary to characterize thermal hysteresis even though its existence is prevalent. The hysteresis may even change as a result of physically handling the unit since the causes of the hysteresis span from characteristic changes of the integrated circuit components to package-induced stresses.

Line regulation is measured by monitoring the changes in the reference while varying the input supply voltage throughout its specified range. As line regulation is evaluated, quiescent-current flow can also be monitored throughout the input voltage range. Since they are steady-state parameters, the variation of the supply should be relatively slow. In other words, a minimum amount of time should be spent at each input voltage setting before data is documented. The bandwidth of the reference determines this limitation, which is defined by the load capacitance and the output impedance of the circuit. For instance, if the bandwidth of the circuit is 20 kHz, then the minimum time that should be spent at each input supply setting is roughly 20 μs (2.3 $\div$ $2\pi f_{BW}$, where f_{BW} is the bandwidth in Hertz and the equation is derived in Appendix C.5). Micropower designs with low quiescent-current flow tend to have lower bandwidths, thereby requiring slower measurements. Similar considerations must be taken into account when gauging any steady-state parameter, like load regulation performance.

For accurate line regulation measurements, the supply voltage should descend from its highest setting to its minimum specified value. This limitation is imposed to prevent *startup operation* and associated delays from affecting the measurement. If the supply voltage were to ascend from zero volts, the circuit will traverse from its zero-current state through startup to its final normal operating mode. Startup occurs when the input voltage is below its minimum specified value. However, the delay associated with startup is typically much larger than the one defined by the bandwidth of the circuit. As a result, if the supply is descended from its maximum specified value, the circuit will not suffer from any startup problems since the measurement ceases before the onset of the zero-current state. On the other hand, the input supply voltage should ascend from zero volts when ascertaining the robustness of the startup capabilities of the reference. Different power-up sequences can also be used as startup experiments (i.e., stepping the input supply from zero to its minimum and maximum specified values within nanoseconds).

Power supply rejection (PSR) can be measured by superimposing a sine wave on the input power supply and monitoring the amplitude of the sinusoidal signal at the reference voltage. The frequency of the sine wave is varied after each measurement to complete acquisition of the PSR performance ($\Delta V_{in} \div \Delta V_{ref}$) for various frequencies. A well-calibrated gain stage may be cascaded to increase the resolution of the measurement system. For instance, if a 0.25 V peak-to-peak voltage were to be used as the AC signal in the input supply of a 45 dB PSR circuit, the expected peak-to-peak voltage of the reference is $(1.4 \text{ mV} = 0.25 \text{ V} \div 10^{45 \text{ dB}/20})$. If the resolution of the equipment is greater than 2 mV, then a well-characterized high-bandwidth 20 dB gain stage can be added such that the actual output signal has a peak-to-peak voltage of 14 mV. As a result, the device under test (DUT) has a PSR performance of 20 dB greater than the actual reading of the measurement, i.e.,

$$\text{PSR}_{\text{db}} = 20 \log_{10} \left(\frac{0.25 \text{ V}_{\text{p-p}}}{14 \text{ mV}_{\text{p-p}}} \right) + 20 \text{ dB} = 45 \text{ dB}. \qquad (5.7)$$

Similar techniques are used when measuring the output noise performance of the circuit. Systematic noise is typically noise injected through the power supplies, thus its performance is described by PSR. Inherent output noise, however, is present even when all injected noise is eliminated. The typical procedure used to characterize this innate output noise begins by characterizing the gain and the noise performance of the measurement system. The resulting noise and gain profiles are eventually used to extrapolate the actual noise performance of the reference circuit. This system characterization should be done just before doing the measurement with the DUT to ensure a consistent testing environment. Consequently, the PCB must be configured to allow the noise of the reference to be shunted to ground when ascertaining the noise contribution of the system. The effects of the noise introduced by the system (V^*_{system}) are subtracted from the total noise (V^*_{total}) by using the sum of squares,

$$V^{*2}_{\text{total}} = V^{*2}_{\text{system}} + V^{*2}_{\text{DUT}} \qquad (5.8)$$

hence

$$V^*_{\text{DUT}} = \sqrt{V^{*2}_{\text{total}} - V^{*2}_{\text{system}}}, \qquad (5.9)$$

where V^*_{DUT} is the noise generated by the device under test (DUT). The accuracy of the measurement increases as the noise introduced by the system decreases. Generally, the system noise contributes excessive noise relative to the DUT if the total noise varies by less than 10% when the noise of the DUT is allowed in the system. Ideally, the noise of the system should be at least an order of magnitude lower than the DUTs. This constraint may be difficult to achieve if the reference itself is low noise. It is noted that the load capacitance of the reference determines the noise bandwidth. In other words, larger output capacitors yield lower noise bandwidths, thereby reducing the total root-mean-squared (RMS) noise of the circuit.

In summary, the measurement of the temperature coefficient must consider *soak time*, *thermal hysteresis*, and *thermal stability* of the surrounding biasing components. Furthermore, in determining line regulation, the input voltage must remain in its steady-state value for at least a predetermined amount of time, which is determined by the bandwidth of the reference circuit. Additionally, the input voltage must descend from its highest setting to its lowest specified setting. The startup quality of the reference can be gauged by subjecting the IC to different power-up sequences such as a slow ramp-up or a sudden step increase of the input voltage. PSR, on the other hand, is measured by superimposing a sinusoidal wave on the input supply and measuring the response of the reference. Cascading a well-characterized high-bandwidth gain buffer and appropriately adjusting the resulting measurement can increase the resolution of the testing environment. Finally, nonsystematic noise is measured in a similar fashion as PSR but using the noise of the DUT as the input. However, the noise contribution of the measurement environment must be low enough to allow the extrapolation of reasonable data.

5.5 SUMMARY

Though many applications still require lower-order references, curvature-corrected circuits are becoming increasingly popular in high-performance systems. The demand for high-performance references stems from several factors ranging from reduced dynamic range (resulting from lower input voltages) to increased system complexity. Constant noise floors and low power supplies, which result from battery operation and finer photolithography, cause the effective dynamic range to be reduced. Low-order references, however, are, and will be, in demand. Given an application whose accuracy requirement is low, a zero-order or first-order circuit is typically the least costly solution. Higher-order references are inherently more complex than lower-order circuits, thereby requiring more silicon area, more quiescent-current flow, and, possibly, higher input voltages.

There are other aspects to designing a reference, besides temperature drift, that are equally crucial to the system and overall product. The output stage, for instance, defines the output impedance of the circuit. In turn, this output characteristic partially dictates the circuits vulnerability to random and systematic noise as well as its ability to source and sink current. The design must mitigate the effects of steady-state changes and systematic noise present in the input supply voltage by designing for power supply rejection (PSR) and line regulation. Cascodes and pseudo-preregulated supplies perform this task well. The design must also account for process variations from die to die, wafer to wafer, and lot to lot. As a result, careful consideration is given to the design of the trim network. Package-shift effects also make this task more difficult. Additionally, the system must be analyzed to determine the effects of interfacing with other circuit blocks. Systematic noise, for instance, may be injected through the substrate, power supply, and/or other circuits, thereby affecting the design of the circuit and the layout. Finally, the parametric limits of a design must be carefully characterized to ascertain and define its dependence to all the different factors, independent of one another.

Most precision references base their operation on the temperature dependence of a diode (base-emitter) voltage. However, the concepts and techniques presented extend to any well-characterized naturally existing voltage. The essence of designing a temperature-independent reference involves the manipulation and the generation of several temperature-dependent components whose sum yields a low TC output. This sum may be accomplished through currents and/or voltages. Furthermore, the techniques used to decrease the susceptibility of the bandgap reference to input power supply, output load, and process variants apply generally to the design of any reference. A very precise reference is achieved when the circuit incorporates all these concepts into its own design.

APPENDIX A.5   BANDGAP TRIMMING PROCEDURE FOR A MIXED-MODE (BOTH VOLTAGE-MODE AND CURRENT-MODE) OUTPUT STAGE

The output structure of a mixed-mode (both voltage-mode and current-mode) curvature-corrected bandgap reference is illustrated in Figure A.5.1. Trimming for the absolute value of the reference voltage is done at room temperature while temperature compensation may be achieved by trimming throughout the specified temperature range. The following procedure illustrates how trimming, in general, can be

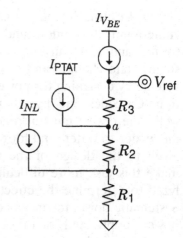

Figure A.5.1 Mixed-mode (both voltage-mode and current-mode) output structure for a reference circuit.

done for this circuit architecture regardless of the temperature coefficients (TCs) of the different components.

- **Step 1:** The reference voltage (V_{ref}) for this topology is expressed as

$$V_{ref} = I_{V_{BE}}(R_1 + R_2 + R_3) + I_{PTAT}(R_1 + R_2) + I_{NL}R_1, \quad (A.5.1)$$

where $I_{V_{BE}}$, I_{PTAT}, and I_{NL} are base-emitter derived, proportional-to-absolute temperature (PTAT), and nonlinear temperature-dependent currents, respectively. Alternatively, V_{ref} can be reexpressed as

$$V_{ref}\left(\frac{R_{2_{initial}}}{R_1 + R_2}\right) = \left[I_{V_{BE}}(R_1 + R_2 + R_3) + I_{PTAT}(R_1 + R_2) + I_{NL}R_1\right]$$

$$\times \left(\frac{R_{2_{initial}}}{R_1 + R_2}\right) \quad (A.5.2)$$

or

$$V_{ref}\left(\frac{R_{2_{initial}}}{R_1 + R_2}\right) = AI_{V_{BE}}R_{2_{initial}} + BI_{PTAT}R_{2_{initial}} + CI_{NL}R_{2_{initial}},$$

$$(A.5.3)$$

where A, B, and C are the coefficients of the temperature-dependent currents,

$$A = \frac{R_1 + R_2 + R_3}{R_1 + R_2}, \tag{A.5.4}$$

$$B = \frac{R_1 + R_2}{R_1 + R_2} = 1, \tag{A.5.5}$$

and

$$C = \frac{R_1}{R_1 + R_2}. \tag{A.5.6}$$

The temperature coefficient of the resistors is canceled because they are ratioed and fabricated with the same type of material, i.e.,

$$\frac{R_1(T)}{R_2(T)} = \frac{R_1(T_r)\left[1 + A(T - T_r) + B(T - T_r)^2\right]}{R_2(T_r)\left[1 + A(T - T_r) + B(T - T_r)^2\right]} = \frac{R_1(T_r)}{R_2(T_r)}, \tag{A.5.7}$$

where $R(T)$ is the value of the resistor at temperature T, A and B are the linear and the quadratic temperature coefficients of the resistor, and T_r is the reference temperature (typically room temperature).

- **Step 2:** Data points are collected for reference voltage V_{ref}, voltage at node a (V_a), and voltage at node b (V_b) throughout the specified temperature range.
- **Step 3:** Since the initial resistor ratios are known and the voltage across each resistor has been collected throughout the temperature range, the currents multiplied by the initial resistance of R_2 are derived for the entire temperature sweep,

$$I_{V_{BE}} R_{2_{initial}} = \frac{V_{ref} - V_a}{\left(\dfrac{R_3}{R_2}\right)_{initial}}, \tag{A.5.8}$$

$$I_{PTAT} R_{2_{initial}} = (V_a - V_b) - I_{V_{BE}} R_{2_{initial}}, \tag{A.5.9}$$

and

$$I_{NL}R_{2_{\text{initial}}} = \frac{V_b}{\left(\dfrac{R_1}{R_2}\right)_{\text{initial}}} - (V_a - V_b). \qquad (A.5.10)$$

- **Step 4:** At this point, the values of the currents multiplied by the initial resistance of R_2 at *room temperature* T_r are identified, i.e., $(I_{V_{BE}}R_{2_{\text{initial}}})_{T_r}$, $(I_{\text{PTAT}}R_{2_{\text{initial}}})_{T_r}$, and $(I_{NL}R_{2_{\text{initial}}})_{T_r}$.
- **Step 5:** Equation (A.5.1) can be reexpressed as

$$V_{\text{ref}}\left(\frac{R_{2_{\text{initial}}}}{R_2}\right) = \left[I_{V_{BE}}(R_1 + R_2 + R_3) + I_{\text{PTAT}}(R_1 + R_2) + I_{\text{PTAT}}R_1\right]$$

$$\times \left(\frac{R_{2_{\text{initial}}}}{R_2}\right) \qquad (A.5.11)$$

and consequently yields the following at room temperature:

$$V_{\text{ref}}(T_r)\left(\frac{R_{2_{\text{initial}}}}{R_2}\right) = (I_{V_{BE}}R_{2_{\text{initial}}})_{T_r}\left(\frac{R_1}{R_2} + 1 + \frac{R_3}{R_2}\right)_{\text{initial}}$$

$$+ (I_{\text{PTAT}}R_{2_{\text{initial}}})_{T_r}\left(\frac{R_1}{R_2} + 1\right)_{\text{initial}}$$

$$+ (I_{\text{PTAT}}R_{2_{\text{initial}}})_{T_r}\left(\frac{R_1}{R_2}\right)_{\text{initial}}, \qquad (A.5.12)$$

or

$$\left(\frac{R_{2_{\text{initial}}}}{R_2}\right) = \frac{(I_{V_{BE}}R_{2_{\text{initial}}})_{T_r}}{V_{\text{ref}}(T_r)}\left(\frac{R_1}{R_2} + 1 + \frac{R_3}{R_2}\right)_{\text{initial}}$$

$$+ \frac{(I_{\text{PTAT}}R_{2_{\text{initial}}})_{T_r}}{V_{\text{ref}}(T_r)}\left(\frac{R_1}{R_2} + 1\right)_{\text{initial}}$$

$$+ \frac{(I_{\text{PTAT}}R_{2_{\text{initial}}})_{T_r}}{V_{\text{ref}}(T_r)}\left(\frac{R_1}{R_2}\right)_{\text{initial}}. \qquad (A.5.13)$$

As a result, the ratio of $R_{2_{\text{initial}}}$ and R_2 is derived for any desired value of $V_{\text{ref}}(T_r)$.

- **Step 6:** The TC of V_{ref} is optimized using equation (A.5.3) (via a spreadsheet) by using the values derived in equations (A.5.8) – (A.5.10) and adjusting the coefficients A and C (B is predefined to be 1) to yield the lowest variation over the specified temperature range. This is an iterative trial-and-error process whose ultimate results are the extracted values for coefficients A and C.

- **Step 7:** At this point, rearranging and solving equations (A.5.4) and (A.5.6) yields new resistor ratios for R_1/R_2 and R_3/R_2,

$$A = \frac{R_1 + R_2 + R_3}{R_1 + R_2} = \frac{\left(\dfrac{R_1}{R_2} + 1 + \dfrac{R_3}{R_2}\right)}{\left(\dfrac{R_1}{R_2} + 1\right)} \quad \text{(A.5.14)}$$

or

$$\frac{R_3}{R_2} = (A - 1)\left(\frac{R_1}{R_2} + 1\right) \quad \text{(A.5.15)}$$

and

$$\frac{R_1}{R_2} = \frac{C}{1 - C}. \quad \text{(A.5.16)}$$

- **Step 8:** Change R_2 by the percentage specified in equation (A.5.13). Resistors R_1 and R_3 are modified according to the relations shown in equations (A.5.15) and (A.5.16), i.e.,

$$\frac{R_1}{R_{1_{\text{initial}}}} = \frac{R_1}{R_2} \cdot \frac{R_2}{R_{2_{\text{initial}}}} \cdot \frac{R_{2_{\text{initial}}}}{R_{1_{\text{initial}}}} \quad \text{(A.5.17)}$$

and

$$\frac{R_3}{R_{3_{\text{initial}}}} = \frac{R_3}{R_2} \cdot \frac{R_2}{R_{2_{\text{initial}}}} \cdot \frac{R_{2_{\text{initial}}}}{R_{3_{\text{initial}}}}, \quad \text{(A.5.18)}$$

where the values for $R_{2_{\text{initial}}}/R_{1_{\text{initial}}}$ and $R_{2_{\text{initial}}}/R_{3_{\text{initial}}}$ are known.

- **Step 9:** If the temperature-drift variation requires further adjustment, then the process is repeated from step 2 onward. An additional iteration may be warranted if the magnitude of the reference is changed significantly from its nontrimmed to trimmed state. This condition arises because the output impedance of the mirrors generating the temperature-dependent currents is finite. If the only parameter that needs adjustment is the magnitude of the reference voltage at room temperature, then the process is still repeated from step 2; however, the temperature sweep and step 4 are skipped while using the values of A and C from the previous iteration. As a result, all the measurements are simply made at room temperature.

NOTES

Knowledge of the absolute value of the resistors is not necessary. Instead, the intrinsic parameters that require control are the ratios of the resistors. Furthermore, if the absolute magnitude is desired to be trimmed at room temperature only, then the procedure is the same except that measurements are only obtained at room temperature and steps 6 and 7 are skipped altogether, thereby relying on simulations for the values of A and C. In fact, once the values of A and C for the particular process technology and circuit have been characterized, only room-temperature trimming is necessary. In the end, this type of trimming algorithm requires at least two resistors to be adjusted, thereby necessitating a relatively complex trim network. As a result of the overall overhead, the aforementioned trimming technique is practical only for a few applications.

APPENDIX B.5   PACKAGE SHIFT-EFFECTS IN BANDGAP REFERENCE CIRCUITS

Compressive stresses, and the resulting strain, can change several physical characteristics of semiconductors. The most significant of these changes are variations in the energy band structure of the semiconductor, which is manifested by increases in minority carrier concentrations. Other important changes include variations in hole and electron mobility as well as generation-recombination levels.

STRESS EFFECTS IN JUNCTION DIODES

Quantum-mechanical studies have shown that changes in the energy band structure result in a decrease in the bandgap energy of semiconductors, like silicon and germanium, and a corresponding increase in minority carrier concentration [5]. Furthermore, these increases in minority carrier concentration depend on the crystalline direction (e.g., [100], [111], or [011]) in which the stress is applied, although this effect is minimal in silicon. The ratio of stressed minority carrier density to unstressed minority carrier density can be written, in general, as

$$\frac{p_n}{p_{no}} \equiv \gamma(\varepsilon_n) \geq 1 \qquad (B.5.1)$$

and

$$\frac{n_p}{n_{po}} \equiv \gamma(\varepsilon_p) \geq 1, \qquad (B.5.2)$$

where p_n and n_p are the stressed minority carrier concentrations, p_{no} and n_{po} are unstressed minority carrier concentrations, and ε_n and ε_p are the position-dependent strains in n-type and p-type material, respectively, which highlight how γ is a function of stress. Figure B.5.1 depicts, theoretically, calculated values of $\gamma(\varepsilon)$ for silicon. The term $\gamma(\varepsilon)$ is a function of several exponentials and could be approximated by a single exponential in the stress region 10^8 to 10^9 Pa, which is the order of magnitude for stresses in plastic packages.

Reverse saturation current I_s, of a p-n junction diode can be written as

$$I_s = \frac{qA\overline{D}_n n_p}{W_B} = \frac{qA\overline{\mu}_n V_T n_p}{W_B}, \qquad (B.5.3)$$

where q is the electric charge, A is the cross-sectional area of the emitter, $\overline{D}_n$ is the average effective electron-diffusion constant ($\overline{D}_n = \overline{\mu}_n V_T$), $\overline{\mu}_n$ is the average effective electron mobility, V_T is thermal voltage, W_B is the base width, and n_p is the base minority carrier density. By substituting equation (B.5.2) in (B.5.3), I_s can be written as

$$I_s = \frac{qA\overline{\mu}_n V_T n_{po} \gamma(\varepsilon)}{W_B} \approx I_{so}\gamma(\varepsilon), \qquad (B.5.4)$$

where I_{so} is the reverse saturation current of the diode in the unstressed state. Compared to increases in minority carrier concentration, stress-related changes in the electron mobility have been assumed to be negligible on I_s, as [5] points out. Thus, one of the results of placing stress on a junction diode is to increase its reverse saturation current.

STRESS EFFECTS ON THE BANDGAP REFERENCE VOLTAGE

The bandgap reference voltage V_{BG} for a first-order Brokaw bandgap reference circuit is given, generally, by

$$V_{BG} = V_{BE} + 2k_c V_T \ln(N), \qquad (B.5.5)$$

where V_{BE} is the base-emitter or p-n junction diode voltage, k_c is a resistor-dependent constant, N is the ratio of base-emitter areas of the two diodes generating the PTAT current in the Brokaw cell ($N = I_{s2}/I_{s1}$), and I_{s1} and I_{s2} are their respective reverse saturation currents. If V_{BGo} is taken to be the unstressed bandgap voltage, the shift in the bandgap voltage (ΔV_{BG}) resulting from package stress is

$$
\begin{aligned}
\Delta V_{BG} &= V_{BG} - V_{BGo} \\
&= \left[V_{BE1} + 2\frac{R_2}{R_1} V_T \ln(N) \right] - \left[V_{BE1o} + 2\frac{R_{2o}}{R_{1o}} V_T \ln(N_o) \right],
\end{aligned}
$$

$$(B.5.6)$$

where the subscript o denotes unstressed values. If the bandgap reference is placed in a uniform stress field and proper layout techniques have been used to match resistors as well as transistors, then the ratio of emitter areas (I_{s2}/I_{s1}) and resistor values remains unchanged:

$$\frac{I_{s2}}{I_{s1}} \approx \frac{I_{s2o}}{I_{s1o}} \Rightarrow N \approx N_o \qquad (B.5.7)$$

and

$$k_c \equiv \frac{R_2}{R_1} \approx \frac{R_{2o}}{R_{1o}}. \tag{B.5.8}$$

Consequently, the bandgap voltage shift becomes

$$\Delta V_{BG} \approx V_{BE1} - V_{BE1o} = V_T \ln\left(\frac{I}{I_{s1}}\right) - V_T \ln\left(\frac{I_o}{I_{s1}}\right) = V_T \ln\left(\frac{I \cdot I_{s1o}}{I_o I_{s1}}\right),$$

$$\tag{B.5.9}$$

where

$$I = \frac{V_T \ln(N)}{R_1}. \tag{B.5.10}$$

Currents I and I_o are the corresponding collector currents of the stressed and unstressed bandgap circuit, respectively. Since N is approximately equal to N_o, and assuming that the stress effects on resistors are negligible (e.g., polysilicon resistors [17]), I is roughly equal to I_o, thereby making package shift ΔV_{BG} equal to

$$\Delta V_{BG} \approx V_T \ln\left(\frac{I_{s1o}}{I_{s1}}\right) \approx V_T \ln\left(\frac{1}{\gamma(\varepsilon)}\right) = -V_T \ln[\gamma(\varepsilon)]. \tag{B.5.11}$$

Therefore, the effect of package stress on the bandgap reference is to lower its voltage by an amount approximately given by equation (B.5.11), which is typically between 3 and 5 mV at room temperature.

TEMPERATURE EFFECTS ON THE BANDGAP'S PACKAGE SHIFT

Theoretically, postpackage stress is a result of a difference in the thermal expansion coefficients of the plastic mold and the silicon die. Therefore, it can be shown, approximately, that:

$$\varepsilon \propto s \propto (T_{sp} - T), \qquad T < T_{sp} \tag{B.5.12}$$

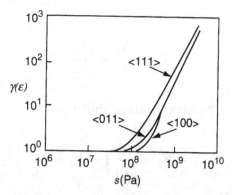

Figure B.5.1 Ratio of stressed to unstressed minority carrier density as a function of stress (s) in silicon. Values are given for $\langle 000 \rangle$, $\langle 111 \rangle$, and $\langle 011 \rangle$ uniaxial compression stress [5].

where ε is the die strain, s is the stress on the die, and T is temperature below the molding set point T_{sp}, which is usually around 175 °C [12]. The stress placed on the die by the plastic is directly proportional to the molding set temperature. As a result, the bandgap voltage shift worsens with higher temperature differentials, from the molding set point. Because stress is directly proportional to temperature deviations, and since $\gamma(\varepsilon)$ can be approximated by an exponential relationship (Figure B.5.1), equations (B.5.11) and (B.5.12) show that the bandgap voltage shift resembles a parabolic relationship with temperature,

$$\Delta V_{BG} \approx -V_T \ln[\gamma(\varepsilon)] \approx -V_T \ln[e^{k_1 s + c_2}]$$

$$= -V_T \ln[e^{c_1(T_{sp} - T) + c_2}] = -c_0 T [c_1(T_{sp} - T) + c_2], \quad \text{(B.5.13)}$$

where k_1, c_0, c_1, and c_2 are physical, semiconductor, and package-related constants.

Figure B.5.2 shows the temperature variation of the bandgap package shift for several devices. Note that the empirical data indicates a mostly linear relationship of package shift versus temperature, which may result simply because a relatively small range of temperature was measured. However, the TC of the package shift shows variation from unit to unit, just like package shift itself. Assuming a mostly linear TC for the package shift, the average TC was about 0.044 mV/°C and its standard deviation was about 0.012 mV/°C. Figure B.5.3 shows a temperature measurement of a bandgap reference before and after packaging. Note that the TC of the bandgap

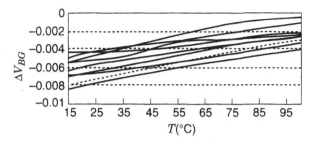

Figure B.5.2   Measurements of package-shift offset (ΔV_{BG}) with respect to temperature.

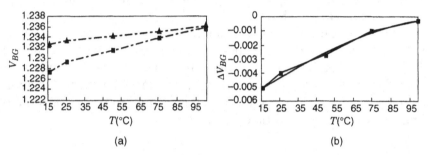

(a)   (b)

Figure B.5.3   Temperature variations of (a) V_{BG} before and after packaging, and (b) ΔV_{BG}.

reference becomes more positive after packaging. The designer, as a result, can compensate the mean by including the TC of the package shift in the temperature compensation of the circuit.

APPENDIX C.5   DERIVATION OF THE TIME REQUIRED FOR A REFERENCE CIRCUIT TO CHANGE A FINITE AMOUNT UPON A SINGLE-STEP STIMULUS

The reference can be modeled by a unity-gain single-pole response system. The model is a first-order approximation. The steady-state change in the reference voltage that results from a variation in the supply voltage is used as the imaginary stimulus for the unity-gain system modeled. This imaginary input has, as the input voltage does, a

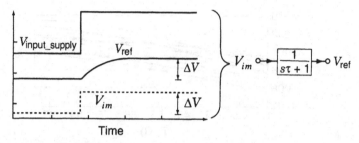

Figure C.5.1 Time-domain description of the imaginary stimulus with respect to the reference and input supply.

rapid step response. Figure C.5.1 graphically illustrates how the imaginary stimulus (V_{im}) is defined. The final response of the new system is equivalent to the one caused by an actual change in the input supply voltage. This manipulation is performed to analyze the delay through the system, or equivalently, the delay that the reference exhibits when a change in the input power supply occurs. As a result, the relationship of the reference voltage (V_{ref}) is expressed as a function of imaginary input V_{im} and time constant τ,

$$V_{ref} = V_{im}\left(\frac{1}{s\tau + 1}\right). \tag{C.5.1}$$

The equivalent input voltage is assumed to have a single-step response with a peak-to-peak voltage of one volt ($V_{im} = 1 \div s$, in the s-domain); thus, V_{ref} is

$$V_{ref} = \frac{1}{s}\left(\frac{1}{s\tau + 1}\right) = \frac{1}{s} - \left(\frac{\tau}{s\tau + 1}\right) = \frac{1}{s} - \left(\frac{1}{s + \dfrac{1}{\tau}}\right) \tag{C.5.2}$$

in the s-domain or, equivalently,

$$V_{ref} = 1 - \exp\left(\frac{-t}{\tau}\right), \tag{C.5.3}$$

in the time-domain, where t refers to time. Consequently, the time that it takes for the reference to rise to 90% ($\Delta t_{90\%}$) of its unity steady-state value (0.9 V) is approximately 2.3τ,

$$V_{\text{ref}}\Big|_0^{0.9\,\text{V}} = \left[1 - \exp\left(\frac{-t}{\tau}\right)\right]\Big|_0^{\Delta t_{90\%}} \tag{C.5.4}$$

or

$$0.9 = 1 - \exp\left(\frac{-\Delta t_{90\%}}{\tau}\right), \tag{C.5.5}$$

which is rearranged to

$$\Delta t_{90\%} = \tau \ln 10 \approx 2.3\tau. \tag{C.5.6}$$

BIBLIOGRAPHY

[1] J.H. Huijsing et al., *Low-Noise, Low-Power, Low-Voltage; Mixed-Mode Design with CAD Tools; Voltage, Current and Time References*. The Netherlands: Kluwer Academic Publishers, 1996.

[2] P.R. Gray and R.G. Meyer, *Analysis and Design of Analog Integrated Circuits*. New York: Wiley, 1993.

[3] A.B. Grebene, *Bipolar and MOS Analog Integrated Circuit Design*. New York: Wiley, 1984.

[4] G. Erdi, "A Precision Trim Technique for Monolithic Analog Circuits," *IEEE Journal of Solid-State Circuits*, vol. SC-10, pp. 412–416, 1975.

[5] J.J. Wortman, J.R. Hauser, and R.M. Burger, "Effects of Mechanical Stress on *p-n* Junction Device Characteristics," *Journal of Applied Physics*, vol. 35, no. 7, pp. 2122–2131, July 1964.

[6] S. Gee, T. Doan, and K. Gilbert, "Stress Related Offset Voltage Shift in a Precision Operational Amplifier," *IEEE ECTC Dig.*, pp. 755–764, 1993

[7] R. Thomas, "Stress-Induced Deformation of Aluminum Metallization in Plastic Molded Semiconductor Devices," *IEEE Transactions on Components, Hybrids, and Manufacturing Technology*, vol. CHMT-8, no. 4, pp. 427–434, December 1985.

[8] H. Miura, A. Nishimura, and S. Kawai, "Structural Effect of IC Plastic Package on Residual Stress in Silicon Chips," *Proc. 40th Electronic Components Technology Conference*, pp. 316–321, 1990.

[9] H. Ali, "Stress-Induced Parametric Shift in Plastic Packaged Devices," *IEEE Trans. Comp., Packag., Manufact. Technology*, vol. 20, pp. 458–462, November 1997.

[10] S.A. Gee, W.F. v.d. Bogert, and V.R. Akylas, "Strain-Gauge Mapping of Die Surface Stresses," *IEEE Transactions on Components, Hybrids, and Manufacturing Technology*, vol. 12, no. 4, pp. 587–592, December 1989.

[11] H. Matsumoto et al., "New Filler-Induced Failure Mechanism in Plastic Encapsulated VLSI Dynamic MOS Memories," *Proc. IEEE International Reliability Physics Symp.*, pp. 180–183, 1981.

[12] H.C.J.M. van Gestel, L. van Gemert, and E. Bagerman, "On-Chip Piezoresistive Stress Measurement and 3D Finite Element Simulations of Plastic DIL 40 Packages Using Different Materials," *Proc. 44th Electronic Components and Technology Conf.*, pp. 124–133, 1993.

[13] T. Regan, "Low Dropout Linear Regulators Improve Automotive and Battery- Powered Systems," *Power Conversion and Intelligent Motion*, pp. 65–69, February 1990.

[14] J. Wong, "A Low-Noise Low Drop-Out Regulator for Portable Equipment," *Power Conversion and Intelligent Motion*, pp. 38–43, May 1990.

[15] M. Ismail and T. Fiez, *Analog VLSI Signal and Information Processing*. New York: McGraw-Hill, 1994.

[16] F. Goodenough, "Fast LDOs and Switchers Provide Sub-5-V Power," *Electronic Design*, vol. 43, no. 18, pp. 65–74, September 5, 1995.

[17] A. Hastings, *The Art of Analog Layout*. New Jersey: Prentice-Hall, Inc., 2001.

INDEX

ABOUT THE AUTHOR

Gabriel A. Rincón-Mora, A Faculty Scholar and a Florida Undergraduate Scholar, earned his B.S. degree in Electrical Engineering from Florida International University, which earned him Honorable Mention by the National Science Foundation. Subsequently, he earned his M.S. and Ph.D. degrees in Electrical Engineering from Georgia Institute of Technology (Georgia Tech), where he was named Outstanding Ph.D. Graduate. He started working for Texas Instruments in 1994 as a Design Engineer, now a Senior Design Engineer, Design Team Leader, and Member of Group Technical Staff. In 1999, he was appointed Adjunct Professor for Georgia Tech. In 2001, he became a full-time member of the faculty in the School of Electrical and Computer Engineering in Georgia Tech, while still working for TI on a more part-time basis. His relationship and work with both TI and Georgia Tech was instrumental in establishing a multi-million dollar sponsorship of the analog program in the School of Electrical Computer Engineering. His work at TI has led to the successful release of many product lines in the field of integrated circuits, power management products in particular, which are already in the marketplace in products like cellular phones, pgers, laptop and desktop computers. He currently serves as a techical advisor for his branch and TI as a whole.

Dr. Rincón-Mora is the author of several journal publications, the inventor of numerous patents, and the designer of many products sold throughout the world. One of his designs was featured on the cover of *Electronic Designs*, a well-respected trade publication, and featured on *EDN's Top 100 Products* for 1998. For his work and contributions to the field of engineering, the Society of Professional Hispanic Engineers (SHPE) honored him with the National Hispanic in Technology Award in early 2000. For his contributions to the field of analog integrated circuit design, he was inducted into Georgia Tech's Council of Outstanding Young Engineering Alumni. He was also among the list of The 100 Most Influential Hispanics, as voted by *Hispanic Business* (a national magazine). For his vision and impact in the engineering field, Florida International University honored him with the Charles E. Perry Visionary Award. The Lieutenant Governor of California also honored him with a Commendation Certificate for his contributions to the field and society as a whole. He has been on the cover of such publications like *La Fluente* (a local magazine owned by the Dallas Morning News), *SHPE* (the Official Magazine of SHPE), and *Hispanic Business*. Additionally, he has also been featured on *EE Times* (a trade magazine) and *Planet Analog* (a web publication), not to mention several other newspapers and publications. He is a Senior Member of the Institute of Electrical and Electtronics Engineers (IEEE), a member of the SPHE, Phi Kappa Phi, Tau Beta Pi, and Eta Kappa Nu.